GW00702314

THE LITERATURE OF GEOGRAPHY

a guide to its organisation and use

THE LITERATURE OF GEOGRAPHY

a guide to its organisation and use

SECOND EDITION

by

J GORDON BREWER

BA ALA MIInfSc

CLIVE BINGLEY
LONDON

LINNET BOOKS
HAMDEN · CONN

FIRST PUBLISHED 1973
THIS REVISED EDITION PUBLISHED 1978 BY
CLIVE BINGLEY LTD 16 PEMBRIDGE ROAD LONDON W11 UK
SIMULTANEOUSLY PUBLISHED IN THE USA BY LINNET BOOKS
AN IMPRINT OF THE SHOE STRING PRESS INC
995 SHERMAN AVENUE HAMDEN CONNECTICUT 06514
SET IN 10 ON 12 POINT PRESS ROMAN BY ALLSET
PRINTED AND BOUND IN THE UK BY REDWOOD BURN LTD
TROWBRIDGE AND ESHER

BINGLEY ISBN: 0-85157-280-4
LINNET ISBN: 0-208-01683-X

Library of Congress Cataloging in Publication Data

Brewer, James Gordon.
The literature of geography.

Includes index.
1. Geography–Bibliography. 2. Geography–Methodology. I. Title.
Z6001.B74 1978 [G116] 016.91 78-16852
ISBN 0-208-01683-X

CONTENTS

FIGURES IN TEXT

INTRODUCTION

> Geography . . . is so comprehensive in character, that the ideally complete geographer would have to know all about every science that has to do with the world, both of nature and of man. (Richard Hartshorne *The nature of geography*)

> There is no problem in this world that is exclusively geographical; but there are few problems that are not in some way geographical. (Saul B Cohen *Problems and trends in American geography*)

THESE TWO quotations exemplify both the reasons for producing this book and the thinking behind its approach to the literature of geography. The broad spectrum of interests represented within geographical studies has naturally resulted in an extensive and diverse literature, intensifying at all levels, the problems (found in any discipline) of finding accurate and up-to-date information quickly and efficiently. The geographer's recourse to literary sources, particularly at the research frontiers of the subject, extends well beyond the traditional bounds of 'geography', creating additional difficulties both for the student and for the librarian. At a more fundamental level, the continuing uncertainty and debate about the precise nature and limits of the subject implies the need for an introduction to basic sources of information. It is the object of this book to fulfil each of these needs, and it is directed at both of the above groups of potential users.

The book is not conceived as a fully comprehensive bibliography; it is doubtful, given the nature of geography, if any such attempt could succeed. It is planned rather as an introductory guide, identifying the most useful, the most significant, and the most authoritative sources within each branch of the subject. Reference works–including comprehensive monographs, textbooks, and handbooks–and bibliographic sources are emphasised throughout, as the means of tracing further more detailed and more advanced literature. The efficient use of such sources presupposes an understanding of the structure of both the subject and the literature; this is stressed particularly in the early chapters, and is an important and recurring theme. It is essential in the use of the later chapters, where the specialised branches of geography are discussed individually. Each of these specialities, in which at a research level considerable use must be made of non-geographical literature, is described with reference to information sources outside of geography, and the citation of individual texts is necessarily very selective. In each field, however, appropriate bibliographies and literature guides are described, as well as dictionaries. This enables the geographer to cope with terminological problems and to search for relevant literature in peripheral subjects. It is clearly not possible within the scope of a single volume to make each of these sections comprehensive, since the range of literature in other subjects which is of potential use to geographers is so great. Published guides to this literature have thus been cited as the only practicable alternative to greatly expanding the size of this volume. A second point which should be stressed is that the specialised chapters are not intended as complete statements independent of materials referred to elsewhere in the book; the literature of these sub-disciplines is a part of the total structure and must be used in conjunction with general sources.

No attempt is made here to describe fully or evaluate in depth every title quoted. The brief comments made are in no sense comparable with reviews, or even abstracts. Just sufficient information is given for positive identification of each item, to enable the reader to appreciate its importance within the structure of the literature, and to assess its likely relevance to his particular needs. The emphasis throughout is on English language and recently published material, and in general it has been possible to incorporate items published up to mid-1977. The effort to ensure currency is a frustrating experience, but I have resisted the temptation to include publications announced but not yet actually available: one such which is likely to become an important work of reference is *International bibliography of geography*, compiled by M Veit and to be published by Verlag Dokumentation of Munich. This will contain about 10,000 titles, mainly German, French, and English, covering all fields of geography.

This second edition of *The literature of geography* retains the structure of the original but has been considerably expanded by the inclusion of over four hundred extra references, nearly all of them published since 1972. It has proved much easier to choose new items for mention than to eliminate titles cited in the previous edition; both tasks however have been approached with caution, and the selection presented now is again perhaps rather conservative. In response to comment on the original edition, a chapter on cartobibliography has been added and greater use made of sample pages to illustrate the format of key reference sources. For permission to include these I am indebted to the following (with the relevant figure shown in parentheses): Munksgaard International Publishers Ltd (3), Prof C D Harris and the Department of Geography, University of Chicago (5 and 12), Geo Abstracts Ltd (6), the Royal Geographical Society (7 and 14), the American Geographical Society (8), the Longman Group Ltd (9a), Edward Arnold Ltd (9b), George Philip and Son Ltd (10), Bowker Publishing Co Ltd (11a and 13), British Library Bibliographic Services Division (11b), the Economist Newspaper Ltd (15), CBD Research Ltd (16), Controller of Her Majesty's Stationery Office (17 and 18), Institut fur Landeskunde, Bad Godesberg (19a), Verlag Dokumentation Saur K G (19b).

I am grateful also for the comments and suggestions of both present and former colleagues, and for those made by reviewers of the first edition. Most of all I am grateful to my family for their patience, and in particular to my wife who assisted in checking and preparing the final draft.

J G Brewer
Bedford College of Higher Education

One

THE LITERATURE OF GEOGRAPHY: SCOPE, STRUCTURE AND USE

GEOGRAPHY is a subject which defies concise definition. From its origins in man's natural curiosity about the world in which he lives–a curiosity as old as man himself–it has become a complex academic discipline embodying a distinct view of the human environment, employing sophisticated techniques of description and analysis, and capable of many applications to social and technological problems. This evolutionary history is to a large extent responsible for the difficulties which inhibit precise definition: the geography of today is different from that of any previous generation. There are, however, certain themes which have remained constant in geographical thinking despite all variations in philosophy and fashion, and definition is best sought in the identification of these.

One of the oldest of these themes concerns the study of the physical nature of the earth and its resources. Several of the branches of modern geography derive directly from this tradition: the relief and form of the earth's surface is the subject of geomorphology; the water features on and near the surface are studied in hydrology and oceanography; the atmosphere surrounding the earth is the concern of meteorology and climatology; and biogeography deals with plant and animal life, and soils. These four areas–the lithosphere, the hydrosphere, the atmosphere, and the biosphere–together constitute physical geography.

The second traditional feature of geographical studies, also apparent from a very early date, is that of the man/land relationship. The placing of man in the context of his environment is a logical development stemming from the physical study of the earth, and from this are derived those branches of the subject collectively termed human geography. Almost any aspect of human activity may be related in this way to the physical character and resources of the earth's surface, resulting in divisions such as economic geography, political geography, settlement geography, and cultural geography. This list is by no means exhaustive, and there is clearly plenty of scope here for the further development of specialised sub-disciplines, a number of which are referred to later in this book, in the chapter on human geography.

Perhaps the most popular and generally accepted view of geography is the idea of area study, and this has in fact long been an integral part of the subject. Ever since exploration and conquest began to extend the frontiers of the known world, the description (and in particular the cartographic representation) of foreign lands has been a central concern of

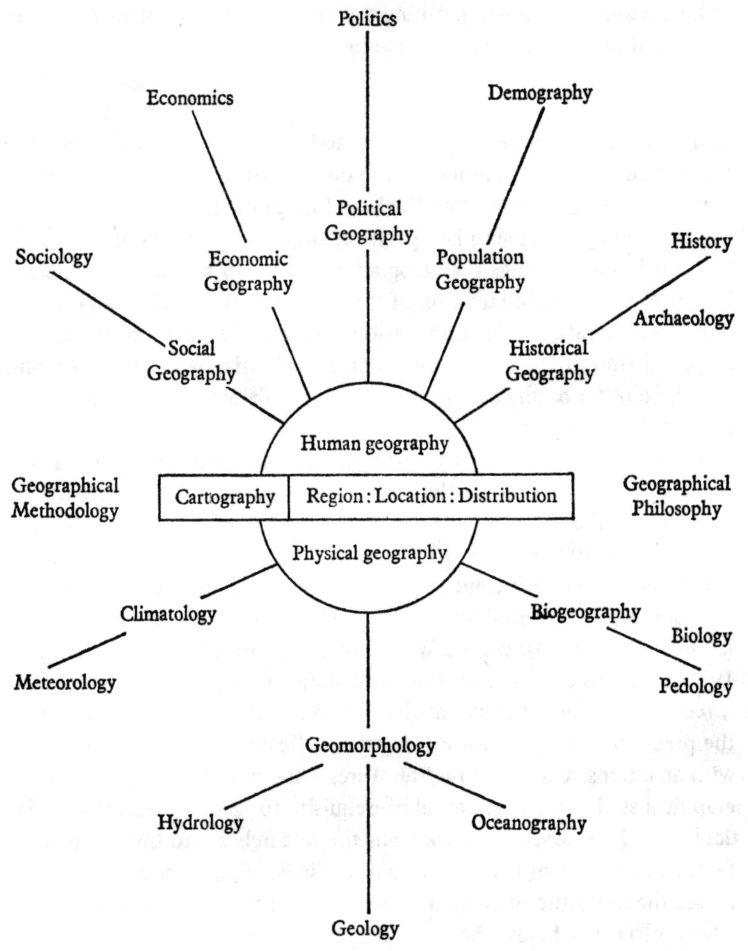

Figure 1: Relationship and context of geographical ideas, methodology, and major sub-disciplines

geography. From this has developed both the synthesising study of regional geography, which looks at both physical and human features in their relationship, and the uniquely geographical methodology of cartography. Further extensions of the regional idea are the concept of geography as the science of spatial relationships (within and between regions); the currently fashionable concern for environment (ie the synthesis of human and physical features in which we live); and recent developments in the use of psychological data to explore man's perception of his surroundings (distinguishing the perceived environment from the real world of objective quantifiable fact).

Scope

The evolutionary development of modern geography from these different traditions is not, however, the only factor which makes definition a problem. It is obvious from the broad range of studies indicated above that geography occupies a key position among other fields of knowledge. The much quoted adage that geography is 'the mother of the sciences' is central to any understanding of the subject, and indeed to an effective use of the literature. The implication is that at the frontiers of geographical study there are numerous areas of overlap with other disciplines and these outer boundaries are as difficult to delineate as the central core of the subject is to define.

Geographers of necessity borrow extensively from these peripheral fields, and research frequently involves an acquaintance with essentially non-geographical literature. For this reason it is important that the geographer understands the way in which information is recorded, and is able to manipulate efficiently the tools of bibliographic organisation. One of the general introductions to the subject (*Geography: an outline for the intending student*, ed W G V Balchin, Routledge and Kegan Paul, 1970) contains a section on the basic skills of geography consisting of three chapters respectively entitled Literacy, Graphicacy, and Numeracy: the precedence given to these is most significant. The ability to cope with an extensive and diffuse literature, while not specifically a geographical skill, is a fundamental prerequisite for geographical study, particularly where advanced work leads the researcher into the possibly unfamiliar literature of neighbouring disciplines. This guide not only discusses the literature of geography itself in some detail, but also provides a lead into other fields where this is appropriate.

The skill of graphicacy referred to above introduces a specialised and unique aspect of the literature of geography. The use of graphic techniques for the presentation and illustration of geographical information has led to the growth of a complete 'literature' composed entirely of maps, plans, charts, and atlases. This material is generally stored and organised in libraries separately from books, journals, and the like (largely for physical reasons), and is listed separately in bibliographies, catalogues, and indexes. Cartobibliography, although strictly speaking an integral branch of the literature of geography, is also therefore a topic of considerable interest in its own right. There is a wealth of material here, making detailed guidance essential if the geographer is to become and remain fully aware of what is available. While it is not possible in a work of this size to provide a fully comprehensive account of this specialised area, chapter six offers a brief introduction to the main sources of information.

The literature of geography is, predictably, international in scope, and the use of source materials in various languages other than English is necessary for advanced work. However, according to a recently published analysis, English is the main language of geographical research (with 77.4 per cent of all items included in the survey), followed by French (8.8 per cent) and German (8.2 per cent). A second finding, perhaps also predictable, and very significant for the study of regional geography, is that geographical publications on any country are overwhelmingly in the official language of that country. A reading knowledge of appropriate languages is obviously therefore an important asset to the geographer, although it is not essential.

There are several sources of information about published translations, many of which are available in the UK from the British Library Lending Division. Failing this, there are directories of translators and translating services from whom translations from any language may be obtained for an appropriate fee. Most academic libraries have a selection of such sources, and the local library should be consulted for details of available translations or of suitable translators.

Channels of communication

The diffuse nature of geographical interests inevitably means that the data used by geographers is generated by a wide range of processes, the number and variety of which have changed considerably as the subject has evolved. It is unwise in these circumstances to attempt a complete rationalisation of the origins of information, as this implies artificial restrictions which are uncharacteristic of the growth and development of the subject. However, since in the primary literature the sources of information are typically reflected by different literary media, some indication of these may be helpful.

During the early pioneering days of modern geography, much information was gathered as a result of individual exploration and discovery, and published in diaries and personal journals and in official accounts of expeditions. Organisations such as the Hakluyt Society have done much valuable work in making this elusive material more readily accessible. Nowadays, the observation and recording of geographical features is more normally a corporate exercise, undertaken by specialist teams using sophisticated research techniques, and publishing the results in reports and surveys, sometimes in cartographic or statistical format. Much work in this category is applied rather than pure geography, and many of the organisations involved are not specifically geographically orientated.

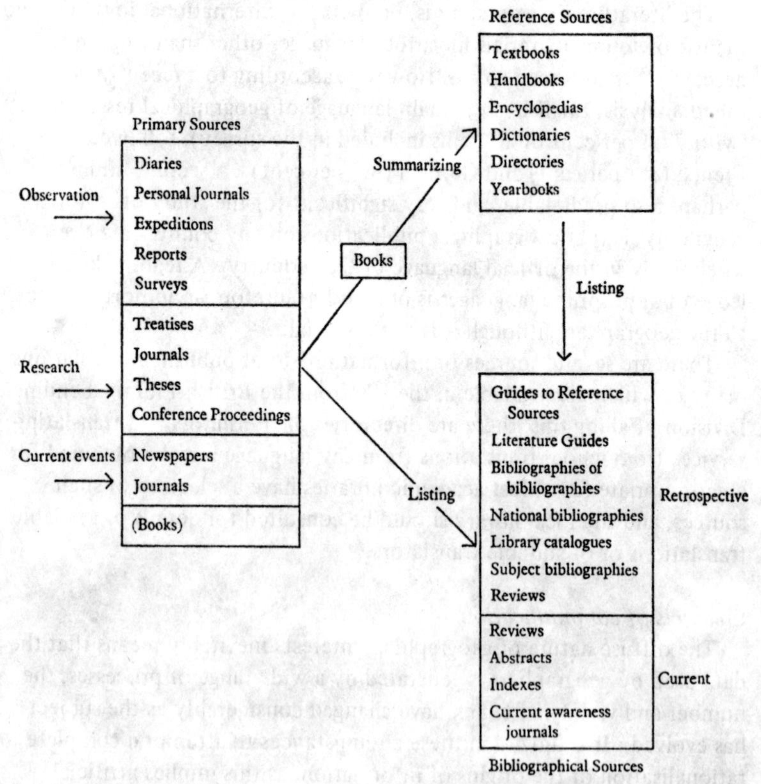

Figure 2: Origins of information, channels of communication, and bibliographic structure

Many government departments and commissions, commercial interests, and professional and research associations contribute to this class of literature.

Another characteristically early form of publication is the treatise, a learned dissertation and exposition, frequently of considerable length, on a particular theme researched by the author. Such works stand out as landmarks in the history of every subject, and geography is no exception. The function of the treatise is quite different from that of the modern book, which is now not so extensively used as a medium for reporting research. Books are not written and published fast enough to fulfil this role any longer, and tend rather to consolidate or distil previously published research results. Books (meaning monographs, textbooks, and the like) are therefore listed among the 'primary sources' in figure two only with reservations.

Most modern research in geography, with the exception of work in applied fields discussed above, has its origins in academic institutions. Such 'pure' research is normally first published in journals, or submitted for degrees in the form of a thesis. These media have the advantage of a kind of inbuilt quality control: articles are normally submitted by journal editors to expert referees before final acceptance for publication; and theses are obviously subject to critical vetting before a higher degree is awarded. Another form of primary publication for original work is the conference paper, which may be either commissioned from an expert or submitted for consideration by a prospective author and subsequently sent to an expert referee for assessment.

Geographers, and particularly those with human and regional interests, are vitally concerned with the changing situations of a developing world. Information which keeps the geographer abreast of significant events is reported in several types of publication, of varying degrees of currency. Initially, newspapers provide the most up to date information, followed by reports in certain journals which specialise in news features. Later, these details are consolidated in reference compilations such as the *Geographical digest* (see chapter three).

This process of summarisation and repetition of information is of course not confined to the reporting of news. Information originally published in most other categories of primary literature is subsequently repeated and reinterpreted in other formats. This is the role of the majority of monographs and textbooks, and although this may be unkindly described as the rehashing of out-dated information, it is usually both more accessible and more digestible in this form. The ultimate stage in the same process is represented by reference works such as encyclopedias and dictionaries. In using sources like this it is advisable always to bear in mind the sometimes considerable lapse of time between publication in handy reference volumes and the original generation of the data and its appearance in the primary literature.

Bibliographic structure

The distillation of information into successively more concise and convenient formats is one of the means by which the sometimes confusing volume and variety of the primary literature is controlled and made more generally available. A second method is by the compilation of bibliographies, whose function is to list and index earlier publications and other documents. Bibliographies themselves conform to no standard pattern: they appear as appendixes to books and other documents, or as separate entities; some attempt to cover a broad area such as the total book production of a country, while others are restricted to one form of literature (eg theses, statistics) or to a narrowly specialised subject field; the amount of information given in each entry may be minimal (just adequate for identification purposes), or include detailed abstracts or reviews (for evaluation); a bibliography may be a single work, perhaps subsequently up-dated by revised editions, or a serial publication providing continuous coverage of the literature as it appears.

Despite the apparently infinite range of variations on this theme, the original simple definition of a bibliography as a list of documentary sources still holds good. The best way to understand the usefulness of all these lists is to see them in the context of their relationships to each other. They are in fact complementary, each serving a slightly differing function, and their effective use by the researcher is governed by an appreciation of their respective places in the bibliographic structure. The major types of bibliographies, which are either interdisciplinary or are represented in most subjects, are discussed below. A knowledge of this overall structure is essential for research in a discipline like geography which relies not only on so-called 'geographical' sources of information, but on comparable sources in related fields.

In figure two, bibliographic sources are divided into two main groups, each of which forms a hierarchical system. The first group, collectively known as retrospective bibliography, is arranged as a hierarchy based on scope. This begins with literature guides, which provide a bird's-eye view of the full range of information sources within particular fields. A similar, though more restricted function is served by bibliographies of bibliographies and guides to reference sources. There is unfortunately no such thing as a universal bibliography, listing comprehensively all that has ever been published; the nearest approach to universality are the national bibliographies, which attempt to list the total published output of books by country. Some large libraries, and libraries having particularly strong collections in certain subject fields, have published catalogues of their holdings. Both these and the unpublished catalogues of less extensive (though perhaps more accessible) libraries may be extremely useful sources of information. Many bibliographies on specific subjects are also available, ranging from monolithic multi-volume compilations of a fairly general nature to short lists of a few items dealing with some specialised research topic.

Current bibliographies are rather more easy to categorise. As the term itself suggests, the intention is to provide up-to-date coverage of the literature as soon as possible after publication. But speed in the reporting of new publications inevitably results in some loss of detail, and it is this balance between currency and the degree of annotation which determines the respective value of the various types of service. Current awareness services, foı example, are useful for indicating the contents of current journals, but have little lasting purpose, since they do not normally cumulate over a period or provide indexes. Indexing journals are of more permanent value and are typically published more slowly and in more durable form. Typically also they are arranged in a classified sequence and produce indexes, and consequently can be used for retrospective searching as well as for current awareness.

Whereas the information given about each publication cited in indexing journals is normally sufficient only for identification purposes, in the case of abstracting services this is expanded to include some indication, of the content and hence the relevance of each item. The preparation of these abstracts obviously involves a greater time-lapse after the appearance of the original information. Taking the process to its logical conclusion are reviews, which are usually published rather less frequently, and survey literature on particular topics over a period, with the intention of noting recent developments in the broader context of historical perspective. The review is thus on the borderline between current and retrospective bibliography and is useful in both spheres. Review articles are of course published in many different journals, but some serials specialise in contributions of this kind.

All these and other possible variations between different bibliographic sources mean that some caution is necessary when using them, and it may be helpful to give some general advice at this stage. The precise terms of reference of any bibliography are, obviously, at the discretion of the author or compiler. These are normally indicated in an introduction in the case of separately published bibliographies, and in lists appended to other works the scope should be clear from the context. The completeness of a bibliography can only be judged in the light of this stated policy, indicating selection policy, level and so on. Similarly, introductory explanations of other factors such as layout, arrangement, and indexing should not be ignored if a bibliography is to be used efficiently. Secondly no bibliography can be relied on to be completely up-to-date. Compilation and publication is a lengthy process and this delay will inevitably see the advent of new information which cannot be incorporated. Currency, therefore, is an important criteria in the interpretation of a bibliography, particularly where in a rapidly developing subject area the most recent information is essential. The corollary of this, at the opposite end of the time scale, is that some bibliographies, especially in secondary works like encyclopedias, may contain items long superseded and which are misleading if taken at face value. Hence the importance to the user of awareness of the date of compilation. Thirdly, all annotations and other indications of content (apart from purely bibliographical notes concerning illustrations, maps, etc) should be treated carefully. Many bibliographies are selective, and selection is a subjective decision. Reviews are also subjective, and in both these cases the user has to assess the authority and reliability of the author. Abstracts as such are rather different: they should consist only of statements summarising or indicating the factual content of the original.

Searching the literature

The control of information from the primary literature by means of a bibliographic structure consisting of bibliographies and reference works is designed for three main purposes. Firstly, it provides a quick guide to specific pieces of information, particularly factual information. Secondly, it enables the researcher to verify publication details of known documents as a prelude to locating and obtaining these for study. Thirdly, it permits comprehensive or selective searches to be made for both retrospective and current literature on any topic. All these activities are illustrated in figure three, which demonstrates the pattern of a theoretically typical literature search. The flow chart should be self-explanatory as a guide to the methodology of searching, but some general considerations need further emphasis.

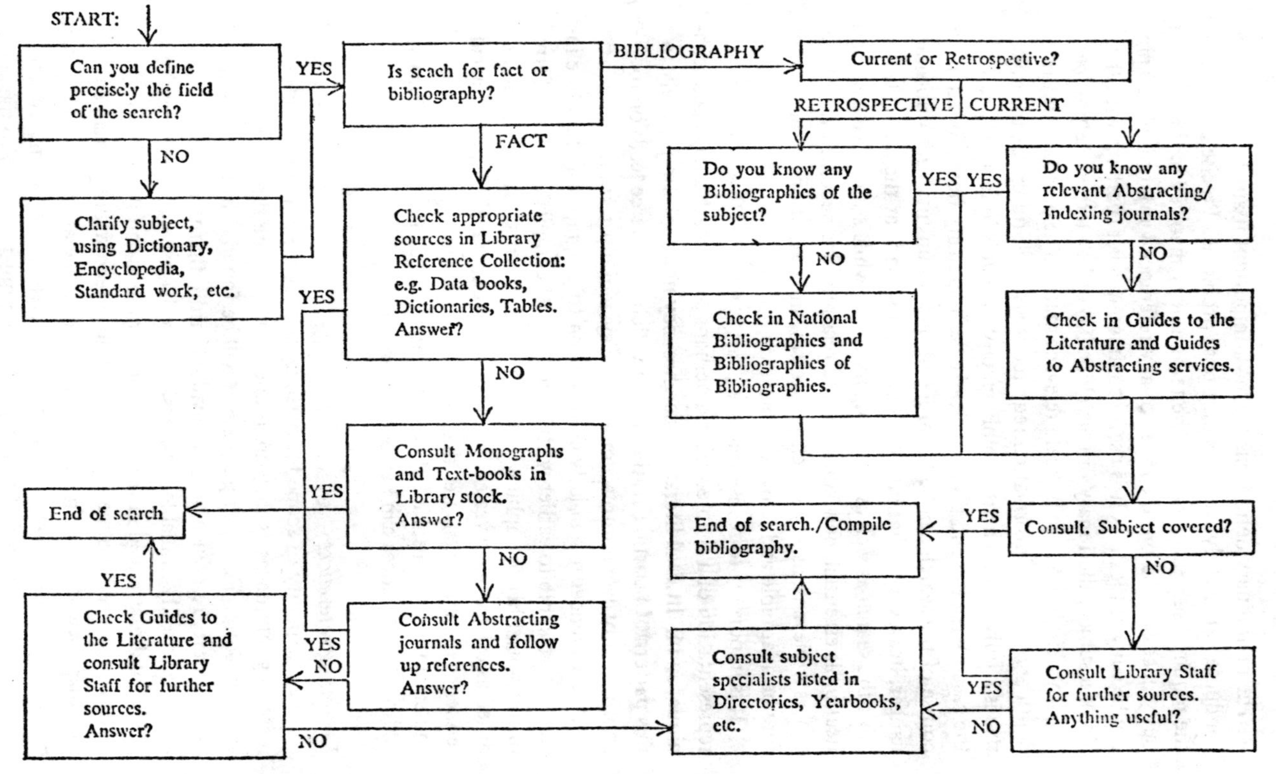

Figure 3: Pattern of a literature search

The preliminary definitions are most important if any search is to be conducted efficiently, and this may pose problems. Precise definition of the required subject scope and the terminology involved in this are obviously necessary before the index of any reference work or bibliography can be fully exploited. It is useful to draw up a list (perhaps with the aid of a subject dictionary or encyclopedia) of key terms likely to be employed in indexes to indicate relevant material. Some idea of the depth and level of information required is also helpful unless the objective is a fully comprehensive bibliography, which is in fact only rarely the case. With the end product clearly defined many items found in the course of the search can be discarded at once as unsuitable. A possible danger, however, is that inflexible definitions made at the outset will inhibit the search as well as clarifying its scope; while some degree of clarification is essential, new concepts and relationships become evident, and others irrelevant, as any search progresses, and definitions should therefore be adaptable and allowed to develop.

It is important to realise that the chart is no more than a schematic representation of a theoretical and perhaps somewhat stereotyped technique, and individual literature searches may not conform precisely to this pattern. In particular, the way in which one source leads on to another is not taken into account in the system illustrated.

Practically every scholarly contribution to geographical literature includes references to earlier writings used in its compilation in the form of footnotes, bibliographies, and acknowledgements. These should be noted as each work revealed by the search is examined. While a logical sequence of sources ensures that coverage is reasonably comprehensive and each item is consulted at an appropriate stage in the search, many additional references can be collected in this way.

Recording references

There is plenty of scope for personal preference in the methods of recording references as they are found, but there are three general points worth making. For most purposes it will be found convenient to note each separate reference on an individual card rather than as part of a list. Cards are readily available in several standard sizes, and these may be arranged and rearranged in the most appropriate order(s). A 5 x 3 inch card is adequate for a merely bibliographical record and is small enough to transport without difficulty provided not too many cards are involved; 6 x 4 inch is more bulky but provides sufficient space for an abstract should this be required. Secondly, attention must be given to the amount of detail recorded about each item, which must obviously be at least the minimum for accurate and unambiguous identification of the publication concerned. For monographs, this normally consists of the author's name, the title, publisher, and date; for periodical articles the author and title are followed by the name of the journal, volume number, date, and pagination. It is well worth ensuring that these essential details are recorded in an appropriate format, particularly if they will eventually be cited in some form of publication; detailed guidance to supplement the outline given above is to be found in the British Standard on bibliographical references, BS 1629: 1976, or in the notes issued by most journals for potential contributors. Thirdly, it is worth incorporating a note of the source of each reference, be it another publication, a library catalogue, or a personal recommendation, unless the writer has actually seen and handled the item. It is fairly frequently necessary in practice to refer back in order to verify details which may be confused, incomplete, or even inaccurate.

These basic rules are sufficient for the construction of a simple record which will prove adequate for most purposes, but it is possible without too much difficulty to create a much more sophisticated index, suitable for holding large numbers of references in a permanent personal information file. These techniques and the simple equipment necessary for some of them, are well described in A C Foskett *A guide to personal indexes* (2nd ed, Bingley; Hamden (Conn), Linnet, 1970), and G Jahoda *Information storage and retrieval systems for individual researchers* (Wiley, 1970). Such indexing systems greatly facilitate the efficient retrieval of all items on a particular subject from a large file containing a variety of material, and in particular are useful when dealing with complex areas covering several related or overlapping subjects or facets of a subject.

Other sources and techniques

The great majority of the sources used in an orthodox literature search of the type described above—indexing and abstracting services and other bibliographies—are arranged by subject and depend for effective use on the ability of the user to define adequately the field of his enquiry. There are, however, some alternative indexing tools known as citation indexes, which work on a rather different principle and which can provide a most useful supplement.

The starting point in using a citation index is not a subject keyword but a reference to a publication you already know. If there is just one reference which is known to be relevant to the search, it can be looked up and the index will list details of other publications in which the original known work is cited. The fact that these further publications are also likely to be relevant is the principle on which the index depends; if in fact this proves to be the case, they may be looked up in their turn, leading to yet more references in which they themselves are cited. Two such indexes, *Science citation index* and *Social science citation index* are widely available, and both contain material of interest to geographers, although the titles are somewhat misleading as a guide to actual content. These indexes offer several advantages over more conventional approaches to the literature, and their interdisciplinary method has much to commend it in a subject where the useful literature overlaps extensively into other, specifically non-geographical areas.

A further possibility which can be considered is that of avoiding the effort of a traditional literature search using printed sources, and paying for a computer search. Many abstracting and indexing services are produced by computer and increasingly the data bases for these are being made available for direct access. In some cases the printed versions have in fact been discontinued, and the only access is via a mechanised search procedure. Many large libraries (particularly academic libraries) will act as agents and arrange for a search to be carried out, and in some it is possible to use a terminal for on-line access, which is a much more efficient way of achieving the best results. These developments are well advanced in both the physical and biological sciences, but have not as yet been applied to any geographical services. However there are several accessible data bases in related disciplines, as well as some broader examples such as those of the two citation indexes mentioned above. This is a field of very rapid change and some complexity, and it will be necessary to seek expert advice from a librarian or information officer experienced in the field. It is only likely to be worthwhile if a comprehensive or very wide ranging search is envisaged, involving those aspects of geography with a heavy dependence on the literature of one of those peripheral subject fields which is already covered by a computer-based facility.

There are circumstances in which the bibliographic tools described here prove inadequate, either because a particular piece of information cannot be readily traced in published sources, or because it is too recently published to be written up. For this reason an information search is not necessarily the same as a literature search. There comes a point when one has to rely on verbal assistance from individual authorities or research groups; such non-literary information sources can usually be identified through directories and yearbooks.

Perhaps the most vital point to stress about literature searching is that it is a complex and skilful operation. Experience improves technique and efficiency, and increases awareness of the range of sources available. Librarians and others who specialise in information and work regularly with all kinds of source materials obviously develop broader knowledge of these and greater expertise in using them than can be acquired by reference to a book such as this. Most of the difficulties which arise in using literature can be resolved by exploiting this accumulated experience, and librarians are always willing to give advice and guidance on sources of information.

Two

THE ORGANISATION OF GEOGRAPHICAL LITERATURE IN LIBRARIES

THE FUNCTION of literature, in its broadest sense, may be summarised as the recording and organisation of information; the function of libraries is the collection and organisation of literature. Libraries provide the usual means of access to the various sources of information described in the previous chapter, and enumerated more specifically in the chapters which follow. Our present purpose is to outline some of the library services available and useful for geographical study, and to discuss the methods by which geographical material is organised in library collections.

Libraries

The literature of geography, as we have seen, is extensive both in terms of its broad subject scope and the sheer volume of current research and publishing. Geography is also a subject which enjoys a fairly general public interest. A consequence of these factors is that geography is represented in a large number of libraries of many kinds, differing greatly in the size and depth of their stock, and in their availability to potential users. It is customary to categorise these into four main groups–national, academic, public, and special–and each of these is described below.

Most countries in the world maintain national libraries. Usually there is a law of copyright, which as well as restricting the reproduction of published material ensures that at least one copy of everything published within the country is deposited in a national collection. In addition, such libraries selectively purchase foreign works, and in the course of time large and unique research collections are developed, which are normally available for reference purposes, though not for borrowing.

In the United Kingdom this function is performed by the British Library Reference Division, comprising the collections of the former British Museum Library and the National Reference Library of Science and Invention. The National Library of Scotland in Edinburgh and the

National Library of Wales in Aberystwyth are also entitled to receive, on request, a copy of any work published in Great Britain. Both of the regional libraries have concentrated on material of local interest and hold strong collections in their respective specialisations.

The Library of Congress (Washington), the Lenin Library (Moscow), and the Bibliothèque Nationale (Paris), are the equivalent institutions in the USA, the USSR, and France. These three, together with the British Library are the largest and best known national libraries. Of particular interest to the geographer is the establishment of national library services, encouraged by international bodies such as Unesco, in many of the developing countries of the world. The problems of obtaining for research original materials emanating from these areas should progressively lessen as these national collections grow and undertake the efficient production of bibliographies to make their resources more accessible.

The second group, academic libraries, comprises the libraries of universities, polytechnics, colleges, and other similar institutions. These of course differ greatly, but are all designed with the needs of a particular academic community in mind. Geography is taught and studied in a large proportion of these establishments, and wherever this is the case supporting library collections must be maintained. Since each of these libraries exists to serve a specific clientele, they do not normally lend to outsiders, but most will allow bona fide students and researchers to use their holdings for reference purposes.

Public libraries are available to all, and again most contain items of geographical interest. Although public libraries, as the term itself implies, have as their main objective a service to the public at large rather than to the researcher, they should not be dismissed by the academic as sources of research material. A point of particular relevance to geographers is the extent to which public libraries have developed collections of local material, often of great value and rarity. The public librarian responsible for such a collection can also be a helpful source of local information and contacts.

The special libraries category includes the libraries of industrial and commercial companies, research organisations, government bodies, and associations and societies of various kinds. Many of these are of considerable interest to the geographer, of which the most obvious examples are the libraries of professional geographical associations such as the Royal Geographical Society and the Association of American Geographers. More specialised aspects of the subject are represented in many other essentially non-geographical libraries, and space does not permit detailed enumeration.

Several directories and guides are available for tracing information about library collections. The *Aslib directory*, edited by B J Wilson and currently in its third edition, is the main source for British Libraries. Volume one, covering science, technology, and commerce, was published in 1968, and volume two, on medicine, the social sciences, and humanities, in 1970. These volumes list libraries alphabetically by place, and a subject index is provided. Although it is a detailed source it is now rather dated, and should be supplemented by *The libraries, museums, and art galleries yearbook* edited by A Brink and D Watkins (Clarke, 1976), which covers both the UK and the Republic of Ireland. It is arranged by place, with a brief indication of the scope and availability of each collection. A further useful volume is *British library resources: a bibliographical guide* by R B Downs (Mansell, 1973), which records all known published library catalogues, directories of libraries, and union lists of holdings. For North America there is *American library directory* (Bowker) which is revised biennially, and *Subject collections: a guide to special book collections in libraries* (fourth edition, Bowker, 1975), edited by L Ash and D Lorenz. The most elaborate guide to North American libraries, however, is the *Directory of special libraries and information centers* edited by M L Young, H C Young, and A T Kruzas (three volumes, Gale Research, 1977). This contains some 1400 entries, with subject, geographical, and personnel indexes and an updating service.

Subject collections in European libraries by R C Lewanski (1965) is a further source worth noting. In addition, there are several international guides which refer to libraries in other countries, for example the *International library directory* (third edition, 1969-70, edited by A P Wales), *World guide to libraries* (sixth edition, Verlag Dokumentation, 1974), and *Major libraries of the world: a selective guide* by C Steele (Bowker, 1976). Another useful title, which includes a wider range of institutions other than libraries, is the *World of learning: directory of the world's universities, colleges, learned societies, libraries, museums, art galleries, and research institutes*, published annually by Europa. Finally, it is worth mentioning here the role of more general directories listing research organisations, academic institutions, societies and associations of all kinds, many of which maintain specialised library and/or information services in addition to their other functions; a number of such directories are discussed in chapter three.

No library can be entirely self-sufficient, even in terms of a limited subject interest. Consequently, cooperative schemes have been established for interlibrary lending, which means that the resources of a local library extend in fact far beyond the stock on its shelves. In this country interlibrary loans are organised through the British Library Lending Division and a network of regional library cooperative groupings. Applications for material not available locally can thus be made through one's own home library, which will channel the request into this national system, exploiting for general use the holdings of many specialised and otherwise inaccessible collections both throughout this country and overseas. In using this service it is advisable to bear in mind that such an operation is necessarily a complex process, and that exact bibliographical details of each item required should be supplied, where possible quoting the source of the reference.

Classification

Most library materials are arranged on shelves in subject sections determined by a systematic scheme of classification. The process of classification presupposes the value of bringing together works on the same topic and relating these subjects physically to other fields which are likely to be of interest to the same body of library users. This practice, however, implies certain problems, and classification can be misleading in some circumstances. In particular, the library must be organised for the benefit of all its users, and individual requirements naturally conflict, so that no arrangement can be universally satisfactory. Secondly, irrespective of what subject categories a classification scheme recognises, other potential groupings must automatically be excluded. Classification therefore is not a particularly efficient tool for retrieving precise information or finding specific books, but serves as a general signposting mechanism for those who like to browse.

Geographers have frequently complained of the inadequacy of established library classifications for the organisation of literature relevant to their needs. This dissatisfaction led to the formation at the 1952 International Geographical Congress in Washington of a Commission on the Classification of Geographical Books and Maps in Libraries, which produced its final report in 1964. Certainly the treatment of geographical books in most general libraries is far from ideal, but the causes of this are to a large extent inherent in the nature of the subject itself.

The content of geography is by no means easy to define, and is consequently difficult to isolate among other disciplines. It is more accurate to describe geography as a point of view than as a given corpus of fact and knowledge. In these circumstances where, in a sense, geography is in the eye of the beholder, it is hardly surprising that geographical literature is usually dispersed in a number of locations through most library collections. Indeed one could argue that there are certain advantages in this, since material of interest to geographers (as distinct from geographical literature) is bound to be scattered in any case. In addition, economic geography (for example) has at least as much in common with some aspects of economics as it does with other branches of geography. It is, however, beyond the scope of this book to explore fully the implications of these ideas; it will suffice to explain the prevailing situation, in which regional geography and general geography are most often grouped as the core of the subject, and the various branches of systematic geography are separated and located as subdivisions of other disciplines.

The second problem, which is uniquely geographical, concerns the distinction between the systematic and regional approaches to the subject and the confusion of these in the literature. Many books embody both elements and may be used in either context, and it is not realistic to expect such titles to be classified according to both principles. Division by region with systematic subdivision is as appropriate for geographical study as systematic classification followed by a regional breakdown. The symbols used in classification schemes to represent subjects are composed of various elements which in fact incorporate both systematic and regional concepts where these are present in one book, and the elements are usually arranged so that the systematic division comes first; in other words, the primary grouping of such books on the shelves is systematic, and regional subdivision is featured within this arrangement. This is true of all subjects, not just geography. Material for the study of particular areas and regions is thus scattered throughout the library, and can be difficult to retrieve.

Major classification schemes

There are many classification schemes in use, including some specially designed to cope with geographical literature and for use in geographical libraries. But specialist schemes do not constitute the major source of difficulty, and it is the organisation of geography in general collections which causes most problems. The most popular systems in use are the Dewey Decimal Classification (DC) and its extension the Universal Decimal Classification (UDC), and the Library of Congress Classification (LC). These are discussed below in terms of their treatment of geography. It should be noted that the examples cited of specific class numbers for particular subjects are not to be taken too literally, since all classifications are subject to variations and adaptations to suit the local requirements of individual libraries.

The Dewey Classification was devised one hundred years ago, and its treatment of geography is largely explained in the light of this fact. Dewey divides knowledge into ten main classes, as follows:

000 Generalities
100 Philosophy and related disciplines
200 Religion
300 The social sciences
400 Languages
500 Pure sciences
600 Technology (applied sciences)
700 The arts
800 Literature
900 General history, geography, etc

Each division thus consists of a minimum of three digits, and is subdivided decimally. For example:

500 Science
550 Earth sciences
551 Physical and dynamic geology
551.2 Plutonic phenomena
551.21 Volcanoes
551.22 Earthquakes
551.23 Fumaroles, hot springs, geysers

Geography, for Dewey, is the study of countries and regions, and systematic geography is largely excluded from the main class for the subject; geography is defined as the 'description and analysis by area of the earth's surface and man's civilisation upon it, not limited to a single discipline or subject . . .' and the user instructed to 'class geographical treatment of a specific discipline or subject with that discipline or subject . . .' Recent editions of DC depart from this tradition only to the extent of permitting systematic branches of the subject to be shelved, optionally, with the rest of geography; this is achieved by the creation of a special number (910.1) for 'topical geography', which may be further subdivided by taking the numbers used for a range of other topics from elsewhere in the schedules and adding them to this base. The result is extremely cumbersome, and most libraries will no doubt retain the earlier–and still preferred–practice.

General geography and regional geography are featured in class 900, and certain other aspects of the subject are assigned fairly distinct places within various other classes: Geomorphology (551.4) and Meteorology (551.5) are included among the Earth sciences; Mathematical geography (526) includes Geodesy, Map projections, Surveying, Photogrammetry, and Navigation. Most systematic geographical subdisciplines are effectively branches of other fields, since in Dewey's view they consist essentially of these other subjects considered by area.

For example:

(a)	330	Economics
	330.9	Economic situation and conditions
	330.91-330.99	Geographical treatment (Economic geography, subdivided by area or country)
(b)	574	Biology
	574.9	Regional and geographical treatment (Biogeography, subdivided by area or country)

This subdivision by areas is represented in both these cases (and in the many other possible examples) by certain standard digits, which remain constant in whatever context they are used:

(a)	330.*945*	Economic geography of Italy
	574.*945*	Biogeography of Italy
(b)	330.*947*	Economic geography of Eastern Europe
	574.*947*	Biogeography of Eastern Europe

It is useful for the geographer to learn to recognise these common subdivisions for countries in which he has an interest.

A further problem to beware of is the confusion in Dewey between geography and history. Firstly, Historical geography (classed at 911) is the one systematic branch of the subject to find a place within the main schedule. Secondly, the basic order of the various groups of topics included in class 900 is not only illogical and extremely unhelpful, but fails to distinguish properly between geography and history:

900	General geography and history
901	Philosophy and theory of history
909	General world history
910	General geography and travel
910.1	Topical geography (optional number; dispersal of various topics into other subject areas preferred)
911	Historical geography
912	Graphic representations of the surface of the earth
913	Geography of and travel in the ancient world
913.031	Archaeology
914-919	Geography of and travel in specific continents, countries, localities in the modern world
920	Biography
929	Genealogy

930	General history of the ancient world
940-990	General history of the modern world

Thirdly, many individual libraries and indeed the *British national bibliography* (see chapter three) demonstrate confusion in their application of these schedules to regional books. A text in regional geography and a history of a particular area seem distinct enough; the distinction between these categories is not always as evident in practice as in theory, however, and this is the cause of a great many problems both to librarians classifying the books and to geographers trying to find them.

The UDC has many features in common with Dewey, including the ten main classes of knowledge, but it introduces greater flexibility and more complex notation. The consistent use of identical symbols representing parts of the earth's surface is retained, and they are easier to identify since they are placed in parentheses following the main number:

574.9(45)	Biogeography of Italy
574.9(47)	Biogeography of Eastern Europe

The UDC also provides an alternative place for biogeography and other systematic studies within the main class for geography. This is achieved by adding the original number (as in the example above) to a prefix which files adjacent to other geographical numbers. The two elements combined in this way are linked by a colon. Thus within the geographical class 910 it is possible to include, for example:

(a)	911.2	Physical and natural geography
	911.2:574.9	Biogeography
	911.2:574.9(45)	Biogeography of Italy
(b)	911.3	Human Geography
	911.3:33	Economic Geography
	911.3:33(45)	Economic Geography of Italy

These numbers are somewhat cumbersome and have the effect merely of reproducing on shelves adjacent to regional geography the identical order which would otherwise exist as part of the biology or economics classes. And use of this device, like the similar provision in the Dewey scheme, is at the discretion of the librarian. This may seem an unduly complicated explanation, but each library will normally have made a policy decision either to centralise or disperse geography, and it is relatively easy from the above outline to establish what the local practice is.

The Library of Congress Classification was developed around the turn of the century as the basis for organising the stock of the American national library collection. The notation symbols consist of a combination of letters and numerals, the main classes being indicated by capitals as follows:

A	General works
B	Philosophy and religion
C-F	History
G	Geography and anthropology
H	Social sciences
J	Political science
K	Law
L	Education
M	Music
N	Fine arts
P	Philology and literature
Q	Science
R	Medicine
S	Agriculture
T	Technology
U-V	Military and naval science
Z	Bibliography

The actual content of the geography class G is interesting in three respects. Firstly, its full title (Geography, anthropology, folklore, manners and customs, recreation) indicates the inclusion of some very non-geographical matter. Secondly, although every other scheme recognises regional geography as one of the more tangible aspects of the subject, this is excluded from the LC. Regional geography is placed in the history classes on the grounds that 'a knowledge of the geography of a country is essential to the understanding of its history'. (It might be noted in passing that this confusion of geography and history occurs also to some extent in both Dewey and UDC: although both schemes have separate schedules for the two subjects many libraries in practice tend to incorporate geographical works into history.) Thirdly, systematic studies are represented only in part in class G, others being scattered among other disciplines. For example, Mathematical geography (GA), Physical geography (GB), and Oceanography (GC) are included, whereas Economic geography and Population geography come in class H, and Political geography in class J.

A further inconvenience for the geographer, by comparison with other arrangements, is that the LC classification numbers are not generally constructed synthetically from common recognisable elements representing different facets of subjects. In particular, the subdivision of subjects by geographical area does not employ uniform digits which can be identified within the complete number. Sometimes the process of regional subdivision is done alphabetically by country, and in other cases numerals are used.

In conclusion, it will be obvious to any geographer that none of these classifications of knowledge treat this subject very kindly, and all of them disperse geographical materials fairly widely throughout most library collections. On the other hand, this dispersal can be turned to advantage in that books quite properly shelved with other subjects may be brought to his notice by juxtaposition with apparently misplaced geography books. Once the situation is understood and accepted, the geographer is enabled to exploit more fully the resources of the collection. In any case, classification is not really all that important; it is worth repeating that classification is not intended to be the means of locating specific books and precise information on a subject. This is the role of library catalogues, which are described below.

Catalogues

Library catalogues in fact may serve several functions, and these determine the amount of information recorded about each item in the library's stock. Even the minimum of detail, however, should be sufficient for precise identification and location within the collection of the documents to which the entries relate. This is the most essential purpose of the catalogue, and it is never advisable to ignore its potential as an alternative to the direct approach to the shelves. In many cases the catalogue will contain far more than this minimal information, and the additional detail can be used to form some assessment of the authority, scope, content, and relevance of the book.

A typical entry is illustrated in figure four, which is arranged in the format one would expect to find in a catalogue consisting of a card file. This is probably the most widely-used method of arranging a catalogue, but alternatives are printed books (rather rare, except for the catalogues of national libraries), loose-leaf folders containing entries on individual slips of paper, and the increasingly popular microfilm formats, usually computer-produced. Whatever physical form the catalogue takes, the content is likely to be broadly similar to the entry illustrated, although this represents a fairly full example and some of the detail could be abbreviated or omitted. This is especially true of the COM (computer output microfilm) catalogue, and there has been a strong trend in recent years in favour of brevity; a short entry is certainly adequate for most purposes and most libraries, and fuller details of particular publications are readily available in, for instance, the major national bibliographies.

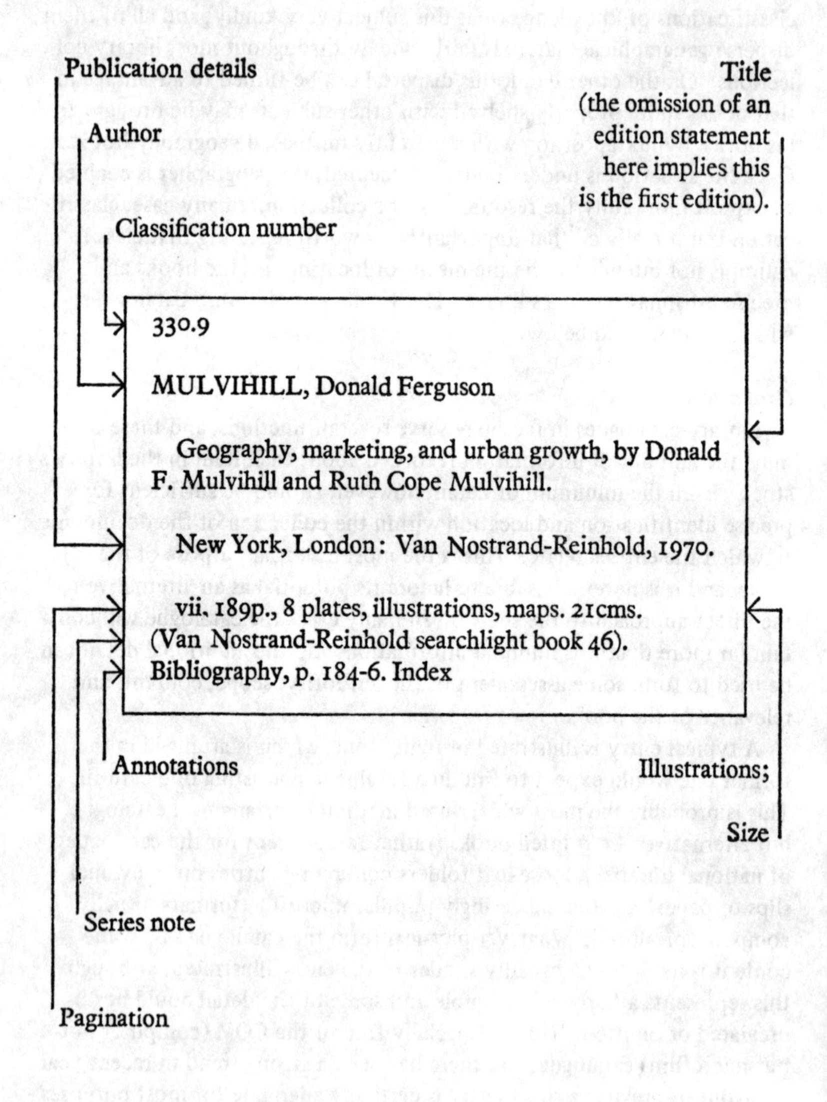

Figure 4: A typical catalogue entry

The entry consists of the name of the author, surname followed by forenames or initials, in the case of personal authors. Symposia and conferences are usually entered under their names, edited works under the names of editors, government publications under the name of the country followed by department or ministry, and institutional publications under the names of the corporate bodies concerned.

For example:

(a) SYMPOSIUM ON CHANGES OF CLIMATE, *Rome*, 1961
 Proceedings of the symposium . . .

(b) STEERS, J A (ed)
 Field studies in the British Isles . . .

(c) GREAT BRITAIN, *Ministry of Transport*
 Roads for the future . . .

(d) INSTITUTE OF BRITISH GEOGRAPHERS
 Land use and resources . . .

In a great many cases, particularly where the main heading is not a personal author, entry is made under more than one of the above possibilities, or under title.

The author is followed by a title statement, the edition (if other than the first edition), the place and date of publication, and the publisher. This information is essential; other details–pagination, illustrations, size, annotations, etc–are susceptible to local variations in practice. Somewhere on the entry, not necessarily at the top, will be the classification symbol, indicating in which part of the library the book is kept. Entries like these are filed alphabetically by the author or other heading to form the *author catalogue*, which is the key to locating specific known books.

Provision is also made for finding out what the library has on particular subjects, and this is by means of similar (or identical) entries in another catalogue, filed in a rather different order. There are two commonly used methods of doing this. The first, favoured by libraries in the USA, is to file alphabetically by subject descriptors added to each entry as headings. For example:

(a) GEOMORPHOLOGY
 SMALL, R J
 The study of landforms . . .

(b) ROADS–Great Britain
 GREAT BRITAIN, *Ministry of Transport.*
 Roads for the future . . .

This type of catalogue also includes many references from one term to another, and these must be followed up if all useful works are to be traced.

Thus:

(a)	SOILS	see	PEDOLOGY
(b)	FROST	see also	PERMAFROST

Sometimes these subject entries and references are interfiled with those under authors, giving a single sequence *dictionary catalogue.*

The alternative system, most common in British libraries, consists of a *classified catalogue*, in which entries are arranged under the classification numbers for each work, thus reflecting the actual order of books on the shelves. This brings together all items at each number and relates subjects to each other according to the structure of the classification scheme in use. This sequence is provided with an alphabetical index from which the letters and/or numbers used to represent each subject can be obtained. Obviously, many terms can occur in more than one context, and index entries are therefore typically given in the following form:

HYDROLOGY	551.48
RIVERS: HYDROLOGY	551.482
RIVERS: TRANSPORT	656.52

By careful use of this index in conjunction with the classified catalogue it is possible to compile a list of the library's holdings on any given topic.

In view of the unsatisfactory classification of geographical literature in most general libraries and the geographer's need to consult material from many separated classes, an ability to make efficient use of these catalogues is essential.

Three

GENERAL GEOGRAPHY: BIBLIOGRAPHIES AND REFERENCE WORKS

A WIDE VARIETY of information sources is discussed in this chapter, including both reference compilations and current and retrospective bibliographies, relating to geography as a whole. It is not realistic, obviously, for the geographer to rely solely on subject-orientated materials, and many general and interdisciplinary reference sources are also described. It will be apparent that although there are close parallels between the several general and geographical works, and between these and similar materials in other subject fields, both have a function within an overall structure and are complementary in many literature searches. Excluded from this chapter are sources confined to sectional interests and specific areas of study within geography, and certain specialised forms of literature such as statistics and government publications.

Literature guides and bibliographies of bibliographies

If the various reference and bibliographic tools by which documented information is controlled are seen as a pyramid with primary literature at its base, the apex consists of literature guides, guides to reference material, and bibliographies of bibliographies. There are a number of brief introductory books of this nature not related to particular subjects, which deal effectively with the best known and most widely used sources of information, and contain hints on tracing books, journals and other records in libraries. Among these are *Facts and how to find them: a guide to sources of information* by W A Bagley (seventh edition, Pitman, 1964), and *How to do library research*, by R B Downs (University of Illinois Press, 1966). Very similar in function and scope is *How to find out: a guide to sources of information for all arranged by the Dewey Decimal Classification* by G Chandler (fourth edition, Pergamon, 1974); this is the introductory volume to a series of subject-based guides, several of which are pertinent to geography and are discussed elsewhere in this book.

At a more advanced level, a number of standard major reference works can be consulted. In contrast to the brief handbooks already referred to, these are less discursive and are designed for consultation rather than reading. One of the best known of these, available in most libraries, is *Guide to reference books* by C M Winchell, to which supplements are issued periodically. This is a single volume work, arranged in broad subject groups, and most of the entries are briefly described. The most recent edition (the ninth) of this well established guide, often referred to simply as 'Winchell', is in fact compiled by E P Sheehy (American Library Association, 1976). The British equivalent is *Guide to reference material* by J A Walford, published by the Library Association. This work, more fully annotated, is in three volumes, covering respectively Science and technology (third edition, 1973), Social and historical sciences, philosophy, and religion (third edition, 1975), and Generalities, languages, the arts and literature (third edition, 1977). General and regional geography form part of volume two, but material of interest to geographers is to be found in all three volumes.

Les sources du travail bibliographique by L-N Malcles (three volumes, Droz, 1950-8 and reprinted in 1965) is another major reference tool of the same kind, although now somewhat out of date. An updated, but much abridged version, entitled *Manuel de bibliographie* (second edition, Presses Universitaires de France, 1969), is also available. A less comprehensive source worth noting for German material is *Handbuch der bibliographischen Nachschlagewerke* by W Totok, R Weitzel, and K H Weimann (fourth edition, Klostermann, 1972). Useful for updating and supplementing the information on reference books in these standard literature guides are the selective lists which appear from time to time in the periodical *College and research libraries* and the *American reference books annual*, published since 1970 by Libraries Unlimited.

There are two guides to the literature of geography which can claim to be reasonably comprehensive, covering a wide range of both general and more specialised reference books, bibliographies and periodicals. *Aids to geographical research* by J K Wright and E T Platt (second edition, 1947) was published by Columbia University Press for the American Geographical Society, and has long been the standard source. Although now obviously dated, it contains many still-useful references to earlier material, and for this reason should not be neglected. Evidence of its continuing relevance is provided by the fact that in 1971 a reprint was issued by Greenwood Press. The successor to Wright and Platt has still not been completed, but *Bibliography of geography* by C D Harris is clearly destined

LOCK, C. B. Muriel. Geography: a reference handbook. 2nd ed. London, Clive Bingley, and Hamden, Connecticut, Linnet Books, 1972. 529 p. (1st ed., 1968. 179 p. 467 entries). Index, p. 429-529. 6

1,283 numbered entries, arranged in alphabetical order mainly by title for books, series, or projects, or by names of institutions or persons. Covers a wide range of geographical material but particularly full on history of exploration and discovery; history of cartography; trade and industrial organizations and their publications, particularly in the United Kingdom; biographical notes on geographers and cartographers; societies and other geographical institutions; atlases and maps; and major substantive publications. Index includes authors, titles, and subjects. Often contains interesting background material about authors, publications, or projects, not readily available in other sources.

NORELL, Irene P. Geographical literature: a brief annotated guide. Compiled under the direction of Irene P. Norell with the assistance of the Department of Librarianship, San José State University, San José, California. San José State University Library, Publication B-3, 1974. 68 p. (Preliminary edition, 1969. 78 p.). Author index. Title index. 7

241 annotated entries organized by types of material: bibliographies and guides to the literature (57); abstracts and indexes (4); encyclopedias, dictionaries, and gazetteers (30); handbooks and directories (34); atlases (77); and history of geography (41). Notes reviews of works listed. Author and title indexes.

FISZHAUT, Gustawa M., and CARRIER, Lois J. Guide to reference materials in geography in the Library of the University of British Columbia. Vancouver, British Columbia, University of British Columbia Library, 1968. 92 p. (Reference publication no. 27). Index. 8

About 500 entries organized systematically, with call numbers in the Library, which uses the Library of Congress classification. Major chapters are general geographic aids, geomorphology, biogeography and climatology, economic geography, social geography, and regions, each subdivided.

FYSH, Patricia V. Geography reference aids in the University of Toronto Library, Humanities and Social Sciences Division. Toronto, Reference Department, University of Toronto Library, 1964. 43 p. (Reference series, no. 9). 9

A useful guide to reference works of geographic value, organized systematically and covering bibliographies, library catalogues, indexes, encyclopedias, almanacs, atlases, dictionaries, directories, gazetteers, government publications, guide books, handbooks, periodicals, statistics, and bibliographies of theses, with annotations. 184 entries.

BURKETT, Jack, ed. Concise guide to the literature of geography. Prepared by students of the School of Librarianship, Ealing Technical College. London, Ealing Technical College, 1967. 47 p. (Ealing Technical College Occasional paper 1/1967). Index. 10

Includes chapters and references: How to find out what exists-- bibliographies; geographical societies; periodicals; reference material; maps and atlases; and storage and retrieval of geographical information.

Figure 5: Sample page from 'Bibliography of geography', by C D Harris

to fulfil this role. Part one ('Introduction to general aids') was published by the Department of Geography at the University of Chicago in 1976, as number 179 in that department's well-known research series. This includes bibliographies, maps and atlases, statistics, photographs, theses, and reference books; each section takes the form of an annotated list of sources, and is introduced by a short general discussion which highlights the main items. A sample page, showing the typical range of the notes provided, is reproduced in figure five.

Among several shorter English language guides which are worth a mention is *Introduction to library research in geography; an instruction manual and short bibliography* by T L Martinson (Scarecrow Press, 1972). In spite of the implication in the title, the bibliography constitutes the greater part of this book and certainly the most useful part, although it suffers from its lack of annotation, descriptive or critical. Another contribution which follows the same two-part pattern is *Geographical research and writing* by R W Durrenberger (Crowell, 1971). In this case the discussion is broadened to include more general information on research methods and technical advice on the preparation of manuscripts for publication. The bibliographies, however, contain an apparently random selection of items with very few annotations. Both of these works reflect a heavy American bias. *How to find out in geography: a guide to current books in English* by C S Minto (Pergamon, 1966) is now badly in need of revision, and the arrangement (following Dewey) is not well suited to the needs of the subject. Two very brief and highly selective sources are *Concise guide to the literature of geography* edited by J Burkett (Ealing Technical College, 1967), and K W Neal *Geography: a guide to geographical publications* (privately printed, 1976).

As the principal English language literature guides are, predictably, biased towards British and American materials, so their counterparts from other countries have emphases which can be very necessary to the geographer. A good example is *Studienbibliographie geographie: bibliographien und nachschlagewerke* by W Josuweit (Franz Steiner Verlag, 1973). Not quite in the same category, but listing similar material for Russian specialists, is *Guide to geographical bibliographies and reference works in Russian or on the Soviet Union*, C D Harris (University of Chicago Department of Geography, 1975). This is a very comprehensive and scholarly guide which is indispensable for advanced work in Russian studies, in the same style as *Bibliography of geography*. For Japan there is *Japanese geography: a guide to Japanese reference and research materials* by R B Hall and T Noh (second edition, University of Michigan Press, 1970).

Although not principally designed as literature guides, there are certain source-books for geography teaching which incorporate much useful information on documentary sources of information in addition to other data. These are discussed in chapter five. In addition many libraries, particularly larger academic libraries with substantial collections of research material, produce their own guides. These have particular advantages in exploiting the resources of the collections on which they are based, but they can also have a wider function; some examples are listed by Harris in the sample page illustrated.

Intermediate between general and geographical literature guides are several guides to the literature of groups of disciplines, including geography. Covering the social sciences as a whole there is *A reader's guide to the social sciences* edited by B F Hoselitz (Free Press, 1967), which includes a concise chapter on geography by N Ginsburg. This chapter deals with physical aspects of the subject in addition to those geographical disciplines which fall more clearly within the social sciences. *Sources of information in the social sciences*, C M White et al (second edition, American Library Association, 1973) contains well-annotated and detailed chapters on the social sciences in general and on nine specialised areas, of which geography is one. Each chapter is contributed by an expert in the field; the geography entries are by C D Harris.

A recent British guide, unfortunately rather less easy to use, is *Use of social sciences literature* edited by N Roberts (Butterworths, 1977). This consists of a series of chapters by several contributors in a narrative style, rather uneven in quality, less self-contained and less rigorously edited than White. More coherent, but less comprehensive, is *How to find out about the social sciences* , G A Burrington (Pergamon, 1975). Individual scientific disciplines are better provided with literature guides than are the social sciences, which is perhaps why more general examples are less in evidence. There is, however *Information sources in science and technology* by C C Parker and R V Turley (Butterworths, 1975), and *Science and technology: an introduction to the literature* by D Grogan (third edition, Bingley; Hamden (Conn), Linnet, 1976). The former is designed as a ready-reference source suitable for those already familiar with the kind of publications available, while the latter contains extensive discussion of the characteristics of these categories of publication, supported by examples.

While guides to reference sources and other literature guides normally include references to and descriptions of major bibliographies, this information in greater detail is also traceable through bibliographies of bibliographies. A concise introductory book in this category is *Bibliographies, subject and national: a guide to their contents, arrangement, and use* by

R L Collison (third edition, Crosby Lockwood, 1968). Moving on to the larger, more comprehensive compilations we have *Index bibliographicus* (fourth edition, Fédération Internationale de Documentation), which consists of two volumes, arranged by UDC, on Science and technology (1959), and the Social sciences (1964). Another major source is *World bibliography of bibliographies and of bibliographical catalogues, calendars, abstracts, digests, indexes and the like* by T Besterman (fourth edition, Geneva Societas Bibliographica, 1965-6). This five volume list is the standard starting point for tracing subject bibliographies, and is now available in more compact subject sections, published by Rowman and Littlefield. Its main disadvantage in use (apart from the fact that, like *Index bibliographicus*, it is dated), is the lack of annotation. To supplement Besterman, which includes material up to 1963, there is *Bibliographic index: a cumulative bibliography of bibliographies* (Wilson, 1937-), a semi-annual publication which cumulates annually. This lists new bibliographies, including those within other works as well as separately published items, in a single alphabetical sequence of titles, authors, and subjects. Also useful for updating the sources cited in the standard guides and bibliographies are the lists which appear in the Unesco periodical *Bibliography, documentation, terminology,* which is available in English, French, Russian, and Spanish editions. These lists are in fact supplements to an earlier Unesco publication, *Bibliographical services throughout the world,* 1965/9 by P Avicenne. The only specifically geographical bibliography of bibliographies as such is *Liste provisoire de bibliographies speciales*, by J Jackson, published in 1881. This lists over five hundred items. More recent bibliographies must be traced through the general bibliographies of bibliographies, through the geographical literature guides and guides to reference sources, and through the current bibliographies.

General bibliographies

The most comprehensive retrospective bibliographies are the catalogues of the world's larger library collections, several of which have been published in book form and are quite widely available. Not only are these useful sources for the verification of bibliographic details of books, and (where a subject approach is provided) as searching tools for the compilation of subject bibliographies, but they do indicate the location of at least one copy of every title. The *Library of Congress catalog* comes in various guises. In its original form it included accessions up to 1942, and a supplement covers the years 1942-6. More recently, the catalogue has assumed an additional role as a current bibliography, and is published monthly, with quarterly, annual and five-year cumulations. This pattern adopted for the author catalogue was in 1950 extended to a quarterly, annual and five-year cumulative subject catalogue. It is not uncommon to find libraries holding some of the cumulations while not subscribing on a regular basis. Another change is in the title, which is now the *National union catalog*; and entries are included which indicate many more locations for the books in major American libraries, in addition to the Library of Congress itself.

The corresponding British catalogue is the *General catalogue of printed books* of the British Museum. In 263 volumes, this is an author list of accessions up to 1955, and annual and ten-year supplements are prepared to keep it up to date. A subject index is also available, mostly in five-yearly sections. For current listing of United Kingdom publications there is a separately published weekly journal, the *British national bibliography* (more usually referred to simply as BNB). This is a classified (Dewey) list with quarterly and annual cumulations, based on books received at the British Library by legal deposit. BNB started in 1950, and use of the annual cumulations for retrospective searching is made considerably easier by five-year cumulated subject indexes. A monthly service, available only on microfiche, which includes a large proportion of the accessions of both the British and American national libraries is *Books in English* (1970-). This, however, is an author list only, and no subject approach is possible.

In addition to these semi-official publications based on the receipts of major libraries, there are numerous trade catalogues which are characteristically less reliable but in some cases more up-to-date. For British books there is *The bookseller*, containing a weekly alphabetical author list which cumulates first into *Books of the month and books to come* (monthly) and then into the quarterly and annual *Whitaker's cumulative book list,* which in addition contains a classified sequence. Similar functions in respect of books in the USA are served by the *Publishers' weekly, American book publishing record,* and *Forthcoming books* (all published by Bowker). Useful for rapid checking of the availability of current books are annual publications such as *British books in print* (now a single alphabetical sequence of authors, titles, and subjects), and the American *Publisher's trade list annual* (a collection of publishers' catalogues) with its indexes *Books in print* and *Subject guide to books in print. Cumulative book index* (H W Wilson Co, eleven issues per year with quarterly and annual cumulations) purports to list books in the English language irrespective of the country of origin; it is particularly strong on American material.

Space precludes description of the many sources in other languages comparable to these English and American publications. It must suffice to note that sources broadly similar in scope are available for the publishing output of other countries, and details may be obtained from the guides to reference sources and bibliographies of bibliographies already noted above.

In addition to these current bibliographies of books, there are a number of general indexes to journal articles which deserve mention. The special value of these to the geographer is that they often include material from fields peripheral to geography (particularly in the social sciences and humanities) which are overlooked by the specialised goegraphical indexing services, and which are otherwise accessible only through similar indexes in other disciplines. *British humanities index* is a quarterly index to about 350 British journals (including important articles from selected newspapers) with an annual cumulation, published by the Library Association. In conjunction with its predecessor, the *Subject index to periodicals*, it covers the period from 1915. In the USA two H W Wilson Co indexes perform a similar function, *The readers' guide to periodical literature* (semi-monthly, 1905-), and *Social sciences and humanities index* (formerly known as *International index to periodicals*, quarterly, 1916-). The former lists material from popular journals, the latter concentrates on the more academic; to quote Walford's example, SSHI indexes *Geographical review*, while the *Readers' guide* covers the *National geographic.* Since mid 1974 SSHI has been published as two separate series, entitled *Social sciences index* and *Humanities index* respectively. An important international index to an extensive range of periodicals in all disciplines is *Internationale Bibliographie der Zeitschriften-Literatur aus allen Gebieten der Forschung* (Osnabruck, Deitrich, semi-annual) which is available (in various formats) since 1897.

Theses are indexed in the *Index to theses accepted for higher degrees in the universities of Great Britain and Ireland*, published by Aslib (the Association of Special Libraries and Information Bureaux). It comprises an annual volume arranging entries within subject groups, with an author index and information on the availability of theses from different universities. North American theses are listed in the monthly *Dissertation abstracts international* and its predecessors, *Dissertation abstracts* and *Microfilm abstracts.* As the titles suggest, each entry is accompanied by a resume of the contents, and the theses listed are available on microfilm from the publishers, University Microfilms. *Dissertation abstracts* is published in two series, on *Humanities and social sciences* (series A) and

Sciences and engineering (series B). The same company publishes *Index to American doctoral dissertations*, which aims to be a complete list, and *Masters' abstracts: abstracts of selected masters' theses on microfilm.* French and German theses are listed in *Catalogue des thèses et écrits académiques* and *Jahresverzeichnis der deutschen Hochschulschriften* respectively.

Report literature is less effectively covered, but is of increasing interest in many fields of geography. *Research and development abstracts journal* (Technology Reports Centre) is published twice monthly and aims to cover all UK government and government sponsored research. The corresponding American service is *Government reports announcements and index* (National Technical Information Service), also twice monthly. Both of these have cumulated indexes and may be used for retrospective searching. Useful for current awareness is the *BLL announcements bulletin,* which is a monthly list of reports, translations, and theses acquired by the British Library Lending Division. The library also produces a valuable monthly guide to conference papers under the title *Index of conference proceedings received by the BLL*, which cumulates annually and supplements *BLL conferences index 1964-1973*; this contains details of nearly fifty thousand conferences, all available from the BLL. This is a difficult area and reports, theses, and conferences are in many cases only semi-published documents; tracing relevant items can be complicated and the above sources are only a selection of the more generally useful. A fuller discussion is available in *Use of reports literature* by C P Auger (Butterworths, 1975), which covers a much broader field than the title suggests.

Geographical bibliographies

The most important retrospective bibliography of geography, and the starting point of any comprehensive search for geographical literature, is the *Research catalogue of the American Geographical Society*, published in fifteen volumes by G K Hall of Boston, 1962. Resembling the catalogue of the Library of Congress, this consists of photocopied entries from the catalogue of the society's library, which represent both books and articles from journals, from 1923 up to 1961. The arrangement is primarily regional, with systematic subdivisions, although the balance is redressed in a supplement (1962-71) currently in progress. This is updated by *Current geographical publications* (see below) and since a complete run of this journal starts in 1938 much of the material prior to 1961 is duplicated, although in a less convenient format. Other library

69C/411 Transport investment strategies and economic development. LEON MONROE COLE, Land Economics, 44(3), 1968, pp.307-317, 27 footnotes.

The main conventional aim of transportation planning is to enable a transportation system to function at minimum cost within accepted safety limits. Reduced transportation costs release capital for investment elsewhere. The author develops a different hierarchy of transportation objectives concentrating on solutions to the social and economic problems of developing countries. Benefits of low-cost transportation accruing to urban commerce and industry are held subsidiary to those of lessening unemployment and promoting training, stabilizing markets for domestic products, controlling the rate of rural-to-urban migration and transforming the rural sector into a money economy. - Russell King.

69C/412 Studies in maritime economics. R. O. GOSS, London, (Cambridge University Press), 1968, 194 pp, 4 figs, 27 tables, 176 refs.

The work consists of a series of collected papers, falling into four sections. First the author undertakes a critical assessment of U.S. maritime regulations pointing out the dangers of unilateral governmental decisions destroying the confidence necessary for the carrying out of international trade, Second, there is a survey of the financial considerations in the production of optimum ship design. The necessity of the comparison of capital return with other capital investment is stressed. Third, there is a comparison of the relative importance of improved ship design, notably containerisation, and of faster port turnrounds. He concludes that the former present greater initial advantages and that the latter should be geared to such improvements. Fourth, there is a study of the complexities of the improvement of port facilities. This is related to the importance of the use of the correct discounted cash flow rate. The conclusion is that once this problem is solved there is a considerable amount of room for port improvement. - Brian Clarke

69C/413 Towards an economic appraisal of port investments. R. O. GOSS, Journal of Transport Economics and Policy, 1(3), 1967, pp. 249-272, fig, 29 refs.

The need for and type of investment in port facilities are discussed but it is stated that in the past there has been no method of appraising investment projects. A method of cost-benefit analysis using calculated shadow prices to represent the social costs and benefits of sea transport is presented. The question of location of investment will need to be answered by a central body as this aspect has international as well as national implications. Practical problems of data availability and collection (for the execution of the appraisal method) and the reform of charging for the use of port facilities (long overdue with or without new investment) are discussed. An appendix is included expressing the method of appraisal in mathematical terms. - Roger K. Lee.

69C/414 Issues and prospects in interurban air transport: a study of socioeconomic, geographic and technical aspects of the development of air travel. B. J. ELLE, Stockholm, (Almqvist & Wiksell), 1968, 230 pp, figs, tables, bibliography.

The prime object of this study is to assess future roles

Figure 6a: A sample page from 'Geo abstracts'

SUBJECT INDEX 1969

MARINE-TERRACES JAVA SUMATRA SULAWESI TERRACES FLUVIAL-TERRACES*	69A/1022
MARINE-TERRACES SANTA-BARBARA-ISLAND(CALIFORNIA) AGE YARMOUTHIAN AFTON	69A/1747
MARINE,* SEDIMENT TRANSPORT- FLUVIAL AND	69A/1797
MARION COUNTIES, SOUTH CAROLINA,*RELATIONSHIP BETWEEN COASTAL LANDFORM	69A/0578
MARION COUNTY, NORTH-CENTRAL FLORIDA,*A POLLEN DIAGRAM FROM MUD LAKE,	69B/1442
MARIOTTE-SIPHON-RESERVOIRS*	69B/1248
MARISEL ALPINE-PASTURES TOURIST-CHALETS HEALTH-RESORTS* CARPATHIANS SE	69D/0690
MARITAL STATUS,* CANADIAN FERTILITY ACCORDING TO RELIGION, ETHNIC GROU	69D/0567
MARITIME ALPS; PRELIMINARY NOTE,*THE HEAD OF THE SIAGNE SUBMARINE CANY	69A/0936
MARITIME CONTINENT' IN THE ATMOSPHERIC CIRCULATION,*ROLE OF A TROPICAL	69B/0811
MARITIME FEDERAL-GOVERNMENT QUEBEC ATLANTIC-PROVINCES BRITISH-COLUMBIA	69C/0670
MARITIME&COMMERCE* PORTUGUESE-ORIGIN IMMIGRANTS FRANCE GERMANY UNITED-	69C/0462
MARITIME-ECONOMICS* U-S SHIP-DESIGN CONTAINERISATION PORT-TURNROUND PO	69C/0412
MARKET?,*WHAT IS A	69C/0642
MARKET ANALYSIS,*GEOGRAPHIC COORDINATES AND	69C/0019
MARKET GARDEN PRODUCTION AND THE PRODUCTION OF DEEPFREEZE PRODUCTS,* D	69C/1017
MARKET HARBOROUGH, (EXPLANATION OF ONE-INCH GEOLOGICAL SHEET 170, NEW	69A/1666
MARKET-AREAS* LOCATION-TECHNIQUES GRAPHIC-MODELS VARIABLE-RESOURCE	69C/0020
MARKET-CENTRE* NATURAL-RESOURCES TASHKENT SAMARKAND BUKHARA	69C/1112
MARKET-GARDEN BARFLEUR LINGREVILLE CREANCES SURTAINVILLE* CAMPAGNE ARG	69C/1170
MARKET-GARDENING STOCK-REARING SIRET DANUBE PRUT FRUIT COVURBI-PLAIN C	69C/1008
MARKET-ORIENTATION SATELLITE-COUNTRY* MARXIST-THEORY COMMUNIST-COUNTRI	69C/0655
MARKET-ORIENTED FISCAL-REDISTRIBUTION LICENSING LEAST-COST-SITE POLITI	69C/0671
MARKET-POTENTIAL DECISION-MAKING BEHAVIOURAL-APPROACH* PARTIAL-EQUILIB	69C/1162
MARKET-POTENTIAL INTERACTION* THRESHOLD SPATIAL-INTERACTION ECONOMIC-H	69C/0965
MARKET-POTENTIALS DISTRIBUTION-CHANNELS PLANT-LOCATION*	69C/0146
MARKET-PRICES* FARM-PRICES PRICE-SUPPORTS FARM-POVERTY	69C/0535
MARKET-RESEARCH HOUSING-ESTATES COMMUTER-VILLAGES LIFE-STYLES SOCIAL-S	69D/1097
MARKET-SEGMENTATION* CONSUMER-NEEDS	69C/0642
MARKET-SEGMENTATION MATHEMATICAL-MODEL* OPTIMAL-ALLOCATION	69C/0645
MARKET-TOWNS BLACK-DEATH WOOL DONCASTER HOPS CANTERBURY KING'S-CLIFFE(	69D/0654
MARKET-VALUE REAL-ESTATE* PERSONAL-VALUE	69C/0718
MARKET,*REGIONAL INCOME DIFFERENCES WITHIN A COMMON	69C/0006
MARKET,*THE DIVISION OF LABOUR IS LIMITED BY THE EXTENT OF THE	69C/0444
MARKETING: THE MANAGEMENT WAY,*	69C/1521
MARKETING* SYSTEMS-APPROACH	69C/0447
MARKETING ANALYSIS MIDWEST* CHICAGO	69C/0019
MARKETING CASH-CROPS* KENYA SWYNNERTON-PLAN(1954) LAND-CONSOLIDATION	69C/0790
MARKETING CHANNEL: A CONCEPTUAL VIEWPOINT,*THE	69C/0438
MARKETING CHANNELS,*SOME OBSERVATIONS ON' STRUCTURAL' FORMATION AND TH	69C/0440
FACTORS GOVERNING THE DEVELOPMENT OF	69C/0441
MARKETING GEOGRAPHER: HIS AREAS OF COMPETENCE,*THE	69C/0144
MARKETING IN NIGERIA,*	69C/0150
MARKETING IN PERU,*THE ENVIRONMENT FOR	69C/0654
MARKETING IN THE INDUSTRIALIZATION OF UNDERDEVELOPED COUNTRIES,*	69C/0145
MARKETING IN THE IRON CURTAIN COUNTRIES,*	69C/0653
MARKETING OF CONSUMER GOODS,*THE ROLE OF GENERALIZATION IN THE	69C/0644
MARKETING ORGANISATION THROUGH THE CHANNEL,*	69C/0439
MARKETING PROCESSES IN DEVELOPING LATIN AMERICAN SOCIETIES,*	69C/1107
MARKETING PURPOSES (1968 EDITION) VOLUME 1,* REGIONAL INDICES FOR	69C/0971
MARKETING RESEARCH POLICIES IN SOUTH AFRICA,*	69C/0914
MARKETING STRUCTURE,* LOWER SAXONY: PRODUCTION AND (AGRICULTURAL)	69D/1106
MARKETING SYSTEM,*INTERACTION OF CHANNEL SELECTION POLICIES IN THE	69C/0448
MARKETING-ANALYSTS MILK-DELIVERY-PROBLEM COMPUTER-TECHNIQUES ROUTE-SIM	69C/0147
MARKETING-CHANNELS TRANSVECTIONS SYSTEM-BEHAVIOUR*	69C/0443
MARKETING-DECISIONS* RESEARCH-AND-DEVELOPMENT INVESTMENT-RISKS	69C/1308
MARKETING-STRATEGIES CASE-STUDIES BRAND-IDENTIFICATION*	
MARKETING,* BRITISH CASES IN	69C/1522
MARKETING,*COMPARATIVE ANALYSIS FOR INTERNATIONAL	69C/1520
MARKETS* INPUT-OUTPUT(MISSOURI) INDUSTRIES MULTIPLIER-RATIO	69C/0659
MARKETS* MANUFACTURING NORTHERN-IRELAND BELFAST-REGIONAL-SURVEY-AND-PL	69C/1063
MARKETS S-NIGERIA* MATRIX VECTOR POPULATION-POTENTIAL ASABA ONITSHA OS	69D/0806
MARKETS,* LONDON'S COMMODITY	69C/0633
MARKHAM AND THE GEOGRAPHICAL DEPARTMENT OF THE INDIA OFFICE, 1867-77,*	69D/0167
MARKIERUNGSVERSUCHE MIT GERUCHSSTOFFEN IN DER KARST-HYDROLOGIE.*	69A/1184
MARKOV CHAIN ANALYSIS TO STRUCTURAL CHANGE IN THE NORTHWEST DAIRY INDU	69C/0277
MARKOV MODELS,* FORTRAN IV COMPUTER PROGRAM FOR SIMULATION OF TRANSGRE	69A/1543
MARKOV PROCESSES AND COMPETITION IN THE BREWING INDUSTRY,* ENTROPY,	69C/0640
MARKOV-CHAIN-MODEL AUCKLAND* REGRESSION NEW-ZEALAND INDUSTRIAL-GROWTH	69C/0241
MARKOV-CHAIN-MODEL WET-DAYS DRY-DAYS*	69B/1019
MARKOV-CHAINS CHI-SQUARE WARD-NEELY PROBABILITY-FUNCTIONS TREND-SURFAC	69C/0005
MARKOV-MODEL BANKFULL-FREQUENCY*	69B/0944
MARKOV-MODEL SACRAMENTO-COUNTY CALIFORNIA SUBURBAN-GROWTH*	69D/0291
MARKSISTKO-LENINSKOE UCENIE O PREVREASCENII SEL'SKOHOZJAJSTVENNOGO TRU	

339

Figure 6b: The computer-produced index to 'Geo abstracts' The marked entry illustrates the choice of index terms used to describe 69C/0412, reproduced opposite catalogues, which may be useful in searching for earlier works, include *Catalogue of the Library of the Royal Geographical Society*, H R Mill (1895), and *Katalog der Bibliothek der Gesellschaft fur Erdkunde zu Berlin*, P Dinse (1903).

Other retrospective bibliographies tend to be more specialised. Useful for tracing basic texts and other important information sources is *A geographical bibliography for American college libraries* by G R Lewthwaite, T E Price, and H A Winters (Association of American Geographers, 1970). As in the book it supersedes (*A basic geographical library*), the American bias is very evident. German geography is selectively listed in *Bibliographie der geographischen Literatur in Deutscher Sprache*, H Arnim (Baden-Baden, Librairie Heitz, 1970), which contains over 1500 unannotated entries, and a bibliography of illustrated geography books is available in *Lexikon geographischer Bildbande* (Vienna, Verlag Bruder Hollinek, 1966). This is international in its scope, but the majority of the references is to regional texts. Geographical theses are listed in *A bibliography of dissertations in geography, 1901-1969*, C E Browning (University of North Carolina Department of Geography, 1970), which covers American and Canadian universities. British university theses are listed regularly in *New geographical literature and maps* (see below), and information in both these sources can be supplemented by use of more general lists such as the *Aslib index* and *Dissertation abstracts.* One other specialised source which deserves a brief mention is *A bibliography of paperback books relating to geography* by H A Hornstein (National Council for Geographic Education, 1970) which lists, under subjects, nearly 650 available titles.

There are several current bibliographies in geography, in various languages, which complement each other quite well. The one most familiar to British geographers is *Geographical abstracts*, published by Geo Abstracts Ltd (from the University of East Anglia). This is unique among the English-language bibliographies in that it is the only one to include abstracts as opposed to simple index entries. Since it first appeared in 1960 it has undergone significant changes in format. Originally entitled *Geomorphological abstracts*, it was restricted to this aspect of the subject until 1965; an annual author index was provided, and this cumulated into an index volume 1960-1965 which also included a subject index. From 1966 the service was extended and the format changed to create four distinct lettered series: (A) Geomorphology; (B) Biogeography and climatology; (C) Economic geography; and (D) Social geography and cartography. Each series produced six issues per year, each was arranged under fairly broad subject headings, and each annual volume was provided with author and regional indexes. An annual index volume cumulated the author indexes of all four series and provided a single subject index. (Cumulative indexes 1966-70 are also available for each of the four series.)

From 1972, there were six series: (B), (C), and (D) remained unchanged, while (A) was renamed Landforms and the quaternary, and two new series were added; (E) Sedimentology, and (F) Regional and community planning. This last series in particular removed an important weakness in the earlier coverage. At the same time the title of the journal was abbreviated to the less formal *Geo abstracts*. Following a large increase in the amount of literature on remote sensing in the early seventies it was found necessary in 1974 to create yet another division, and series (G) on Remote sensing and cartography was begun. This transferred cartography from its former (and slightly illogical) position in series (D), which was accordingly renamed Social and historical geography; historical geography had previously been a minor subdivision within this series. Each of the current seven series produces author and regional indexes, and annual subject and cumulated author indexes to the whole set are also available. The increasing scope of *Geo abstracts* and its monopoly as an English language abstracting journal make this perhaps the most important tool of current bibliography in geography.

Unfortunately however, *Geo abstracts* is a relative newcomer by comparison with some of the well-established indexing services in geography. *New geographical literature and maps* (two issues a year, 1951-) superseded the earlier *Recent geographical literature, maps, and photographs added to the society's collections*, which was published as a supplement to the *Geographical journal.* Entry is made under brief subject descriptors, arranged alphabetically within a regional classification. No index is provided and consequently retrospective searching is extremely difficult. A more useful source is the American *Current geographical publications.* This is a record of publications added to the library of the American Geographical Society, and supplements the published catalogue of the library's holdings discussed already. There are ten issues per year, each consisting of a classified sequence in which regional division predominates, with subject headings interposed. The classification is based on that used in the library. Beginning with the September 1972 issue the classified arrangement is to be modified, placing greater emphasis on systematic (topical) rather than regional divisions. This is consistent with the changes made in the classification used in the ten-year supplement to the *AGS*

BERING SEA CARTOGRAPHY 79

Bering Sea Chirikov Basin Physiography **Hopkins and others**

Geol. Surv. Prof. Papers [Washington] **759-B** (1976), pp. 7

Physiographic subdivisions of the Chirikov Basin, northern Bering Sea. By D. M. Hopkins, C. H. Nelson, R. B. Perry and Tau Rho Alpha. *Maps, bibliogr.*

Greenland Geology Essays **Escher and Watt**

Geology of Greenland. Edited by Arthur Escher and W. Stuart Watt. Copenhagen: Geological Survey of Greenland, 1976. 603 p. *Maps, tab., diagr., ill., bibliogr., ind.* 30 cm. ISBN 87 980404 0 5 £18.50

21 contributions by various authors.

Greenland North-east Hydrology **Allen and others**

Arctic and Alpine Res. [Boulder, Colorado] **8** (1976) 3: 297–317

The chemical and isotopic characteristics of some northeast Greenland surface and pingo waters. By C. R. Allen, R. M. G. O'Brien and S. M. F. Sheppard. *Sketch-maps, tab., diagr., ill., bibliogr.*

Spitsbergen Geomorphology **Wirthmann**

Z. Geomorphol. [Berlin] N.F. **20** (1976) 4: 391–404

Reliefgenerationen im unvergletscherten Polargebiet. By Alfred Wirthmann. *Ill., bibliogr.* [*Engl. sum.*, pp. 391–2]

Svalbard Geography **Hisdal**

Geography of Svalbard: a short survey. By Vidar Hisdal. Oslo: Norsk Polarinstitutt, 1976. 73 p. *Maps, tab., ill., bibliogr.* 19.5 cm. (*Polarhandbok* Nr. 2)

CARTOGRAPHY AND MATHEMATICAL GEOGRAPHY

Air survey Remote sensing **Barrett and Curtis**

Introduction to environmental remote sensing. By E. C. Barrett and L. F. Curtis. London: Chapman and Hall, 1976. xi, 336 p. *Maps, tab., diagr., ill., bibliogr., ind.* 24.5 cm. (Science Paperbacks) ISBN 0 412 12920 5 £11

Air survey Remote sensing **Harper**

Eye in the sky: introduction to remote sensing. By Dorothy Harper. Montreal: Multiscience Publications, in association with the Department of Energy, Mines and Resources and Information Canada, 1976. viii, 164 p. *Maps, tab., diagr., ill., bibliogr., ind.* 20 cm. (Canada Science Series) ISBN 0 919868 01 0 $5.70

Air survey Remote sensing UK research **Department of Industry**

Remote sensing of earth resources. List of UK groups and individuals engaged in remote sensing with a brief account of their activities and facilities. 3rd edn. London: Department of Industry, 1976. v, 266 p. *Tab., bibliogr., ind.* 30 cm. £2

Figure 7: Sample page from 'New geographical literature and maps'

research catalogue, referred to above. The annual index is in two parts: subject, and author and regional. The subject index is not easy to use, as will be seen from the illustration in figure eight: the terminology used is imprecise and subjects are in classified rather than alphabetical order.

Bibliographie géographique internationale was described by Wright and Platt as 'the most convenient, comprehensive, and in many respects the best of all current geographical bibliographies'. Published under the auspices of the International Geographical Union and with the support of Unesco, the BGI is certainly one of the most respected sources in this category, and is the longest established of the indexes still being published, having appeared, normally annually, since 1891. The arrangement is classified, mostly by regions, and there is an author index. A subject approach is possible only by means of a contents list setting out the main groupings, and the lack of a detailed subject index is a serious disadvantage.

Less familiar to many geographers is *Documentatio geographica* (1966-) published by the Institut fur Landeskunde. This appears initially as a bimonthly current awareness bulletin of recent publications, cumulating annually into a two-volume list organised for retrospective searching. One volume contains a classified list of publications cited during the year, the other consists of indexes. There are author and alphabetical subject indexes, and a classified index of the UDC classification symbols under which the entries in volume one are arranged. This includes not only the main UDC numbers but separate entries under the various elements of which the main number is composed. (See chapter two.)

Referativnii Zhurnal is a major international abstracting journal with an emphasis, as one would anticipate, on Russian and East European material. It appears as a number of series, one of which is devoted to geography. This form of organisation is similar to the French monthly *Bulletin signalétique*, although in this latter case there is no specifically geographical series. The range of subject series within both services is considerable (lists are available in Winchell), and is a salutory reminder for the geographer of the several possible headings under which relevant literature may be cited.

A rather different kind of bibliographical source, but one which contains much useful information, is the *Special Libraries Association, Geography and Map Division bulletin.* Published quarterly since 1947, the bulletin contains bibliographical articles and articles on source materials as well as lists and reviews of important new books, maps and atlases.

- 179 -

Nr. 4, 5 and 6. July-December, 1954. p. 143-160. Maps.) Contents: I. Geological research in Timor; an introduction, p. 1-8. - II. The orogenic main phase in Timor, p. 9-20. - III. An occurrence of Miogypsina (Miogypsinella) complanata Schlumberger in the Lalan Asu area, Timor, by P. Marks, p. 78-80. - IV. Notes on Plio-Pleistocene corals of Timor, by R. Osberger, p. 80-82. - V. Structural development of the crystalline schists in Timor, tectonics of the Lalan Asu Massif, p. 143-153. - VI. The second geological Timor expedition; preliminary results, p. 154-160. 42

AUSTRALASIA

O'Connell, D. P. Sedentary fisheries and the Australian continental shelf. (The American journal of international law. Vol. 49. No. 2. April, 1955. p. 185-209.) 45 & 53

Lewis, J. N., & Saxon, E. A. Agricultural output requirements for future population growth in Australia. (Research Group for European Migration Problems /R.E.M.P./ bulletin. Vol. 3. No. 1. January-March, 1955. p. 9-15.) 52 & 53

Borrie, W. D. Economic and demographic aspects of post-war immigration to Australia. (Research Group for European Migration Problems /R. E. M. P./ Bulletin, Vol. 3. No. 1. January/March, 1955. p. 1-8.) 52

Fenner, Charles. Aboriginal place names. (Walkabout. Vol. 21. No. 3. March, 1955. p. 41-42.) 55

Broadbent, H. F. A trip through the North-West (Western Australia). (Walkabout. Vol. 21. No. 3. March, 1955. p. 13-20. Map.) 2

Miles, Thomas A. Tasmania's wild west coast. (Walkabout. Vol. 21. No. 3. March, 1955. p. 29-33. Map.) 11

Pownall, L. L., & Chapman, June E. The location of factory industry in New Zealand. [Wellington, Govt. Printer, 1954] 31 p. Maps. (Post-primary school bulletin. Vol. 8. No. 4.) 53

Gibbs, H. S., & others. Soils and agriculture of Matakaoa County, New Zealand. [Wellington] 1954. 51 p. Maps. (New Zealand. Dept. of Scientific and Industrial Research. Soil Bureau. Bulletin 11.) 43 & 53

PACIFIC ISLANDS

U.S. Board on Geographic Names. Decisions on names in the Trust Territory of the Pacific islands and Guam. Part III: Mariana Islands and Guam. Washington, D.C., 1955. 43 p. (Cumulative decision list, no. 5503.) 55

Macdonald, G. A.; Davis, D. A.; Cox, D. C. Kauai, an ancient Hawaiian volcano. (The Volcano letter. No. 526. October-December, 1954. p. 1-3. Map.) 43

Sherman, G. Donald. Some of the mineral resources of the Hawaiian Islands. Honolulu, 1954. 28 p. Maps. (Hawaii. University. Agricultural Experiment Station. Special publication, no. 1.) 53

U.S. Board on Geographic Names. Decisions on names in the Trust Territory of the Pacific Islands and Guam. Part II: Marshall Islands. Washington, D.C., 1955. 65 p. (Cumulative decision list, no. 5502.) 55

Figure 8a: A sample page from 'Current geographical publications'. The numbers on the right are classification symbols

Figure 8b: Part of the subject index from 'Current geographical publications. The asterisked entry appears opposite

One of the most significant gaps in the bibliographical structure of geography until recently was the lack of a good review journal. The annual series *Progress in geography* has now established itself as such, and increased considerably in size, necessitating a split into two sections. Each contains a selection of excellent reviews with extensive bibliographies, all written by experts closely and currently involved in research activity in their respective branches of the subject. *Geographisches Jahrbuch* fulfilled such a role from 1886, but ceased production during the Second World War. Subsequently revived, it has not succeeded in re-establishing itself as a regular publication, and no issues have appeared since 1967. *Geographisches Jahrbuch* remains, however, a most important source for earlier material and for retrospective searching. Usually annual, each volume contains reviews of the current state of knowledge in various fields of geographical study, with numerous references to the literature. Different subjects are reviewed at varying recurring intervals, and Platt and Wright give a breakdown of these up to 1943, the first part of Band 58. Harris in his *Bibliography of geography* provides a detailed analysis both of subsequent volumes of *Geographisches Jahrbuch* and of *Progress in geography* up to volume 7, 1975.

Bibliotheca geographica: Jahrsbibliographie der geographischen Literatur was another German bibliography of a rather different type. Published 1895-1917 (for the years 1891-1912) it superseded the annual bibliographical summaries previously contained in the *Zeitschrift der Gesellschaft fur Erdkunde zu Berlin*, rather as the *Bibliographie geographique internationale* began life as part of the *Annales de geographie*. Unlike the *Geographisches Jahrbuch*, however, it was a simple, classified list, without annotation or comment.

It is important to stress before leaving the subject of current geographical bibliography that the several series described here are not alternative, but complementary, sources of information. Although international in scope, they all incorporate some degree of bias, and particularly national bias. For example, a recent survey of periodical literature in geography, geomorphology, and geology carried out at the School of Environmental Sciences, University of East Anglia, for the Office for Scientific and Technical Information, pointed out a contrast between *Documentatio geographica* and *Referativnii zhurnal*: in the list of periodicals indexed by the former are sixty-two German titles and only seven Russian: in the latter there are nineteen German and thirty-two Russian titles. Such a bias is of course predictable, but it does emphasise the interdependence of the various indexing and abstracting services in the compilation of comprehensive bibliographical information.

Encyclopedias and dictionaries

It is convenient to consider these two types of reference book together, since in practice it is difficult to assign certain of these works to either category. Although linguistic dictionaries have a quite different function from defining dictionaries and encyclopedias containing explanations and definitions of terms the phraseology employed in the titles of these works can be misleading.

There are no genuine encyclopedias in the English language devoted to geography as a whole, although there is the five volume *Kratkaia geograficheskaia entsiklopediia* (Moskow, Sovetskaia Entsiklopediia, (1960-6), and the German *Westermann lexikon der geographie*, also in five volumes, edited by W Tietze (Westermann, 1968-72). This is a scholarly work with contributions by leading authorities in Germany–some 20,000 entries in all. Many place names are included, but there are also definitions and explanations of technical terms, as well as more substantial articles with quite extensive bibliographies. The standard general encyclopedias may also contain useful introductory information. Among the more important of these are *Encyclopaedia Britannica, Chambers's encyclopaedia, Encyclopedia Americana, Grand Larousse encyclopedique,* and *Der grosse Brockhaus.* Care must be taken in using sources of this nature; not only do they inevitably take a considerable time to compile, thus dating the information contained, but it cannot be assumed that in each successive edition of a major encyclopedia all the articles are revised.

The most scholarly of the several guides to the terminology of geography is *A glossary of geographical terms* (second edition, Longmans, 1966), edited by L D Stamp and prepared for a committee of the British Association for the Advancement of Science. This contains alternative meanings and references to both origins and current usage, sometimes illustrated by quotations, and the sources of information are indicated. Appendixes list stratigraphical terms, foreign words, and prefixes. Based on the material compiled for the glossary is another useful source, also edited by Stamp, *Longman's dictionary of geography* (Longmans, 1966). In addition to technical terms, this includes some place names and biographical entries, and a short bibliography of standard texts from which further amplification may be obtained. *A dictionary of geography*, F J Monkhouse (second edition, Arnold, 1970), is less comprehensive and noticeably stronger on physical than human geography, but its great advantage over Stamp is the way in which the text is backed up by clear, simple diagrams and illustrations. The selection of entries for inclusion is based on a survey of terminology currently in use in the literature, and a list of terms arranged in related subject groups is a useful supplement

loess will probably alter completely into some type of brown loam.' Much of the European 'Lehm' appears to be of such origin (Mill, H. R., *Am. J. Sci.*, 3rd Ser. V., 49, 1895, p. 26). (G.T.W.)

Logan Stone

Mill, *Dict.* A rocking-stone (S.W. England).

Comment. Actually a well-known block of granite in the Land's End peninsula resulting from the atmospheric weathering which produces tors (*q.v.*) which is so balanced that it can be rocked with the hand; applied to similar balanced or 'logging stones' elsewhere.

Loma (Spanish America)

Webster. A broad-topped hill. *Southwestern U.S.*

Dict. Am. A small hill or elevation. Often in place-names. *S.W. U.S.*, 1849.

Mill, *Dict.* A hill or rising ground on a plain (Spanish America).

No reference in standard works, but given with the same meaning by Knox, 1904.

Longitude

O.E.D. Distance east or west on the earth's surface, measured by the angle which the meridian of a particular place makes with some standard meridian.

The meridian which is now almost universally used is that of Greenwich, London, which is considered as 0° and distance is recorded in degrees, minutes and seconds East and West Longitude. In time, fifteen degrees is equivalent to a difference of one hour in local time.

Longitudinal Coast

A coast running broadly parallel to the main tectonic structure or fold lines; also called a Pacific coast being found commonly around the Pacific Ocean. In contrast to Atlantic or Transverse coast.

Longitudinal valley

Mill, *Dict.* A valley whose sides are formed by parallel mountain or hill ranges.

Powell, 1875. '... having a direction the same as the strike.' (p. 160) May be anticlinal, synclinal or monoclinal, *q.v.*

Comment. It should be used in the sense given by Powell. Mill's definition is derived from an earlier use by W. D. Conybeare and W. Phillips, 1822, *Outlines of the Geology of England and Wales*, p. xxiv. According to current usage it is really only correct where the mountain or hill ranges are parallel to the strike. See also Linton, D. L., 1956, The Sussex Rivers, *Geography*, 41, 233–247.

Lōō (Indo-Pakistan: Urdu), **Look** (Sindhi)

Hot dry wind; a heat-wave.

Spate, 1954. 'A very hot dust-laden wind which may blow for days on end' (p. 55).

Loop District

Webster. The Loop. That part of the business center of the city of Chicago surrounded by a 'loop' of elevated railways near the lake and bounded by Lake, Wells, and Van Buren streets and Wabash Avenue.

Dict. Am. Loop. Originally any completed turn in railroad or elevated tracks, then the territory inclosed by such tracks. Especially the territory in downtown Chicago bounded by the elevated railway; hence transferred (especially by Chicagoans), any business district. 1893.

The heart of the central business district of Chicago is approximately delimited by a loop of the elevated railway and hence is known familiarly as the Loop District. Being so well known, Americans not infrequently make reference in other cities to the heart of a CBD (*q.v.*) as the 'loop district'. (L.D.S.)

Lopolith

Holmes, 1944. 'Intrusions, which on the whole are concordant and have a saucer-like form, are distinguished as *lopoliths* (Gr. *lopas*, a shallow basin). The best known examples are of extraordinary dimensions, the largest, that of the Bushveld in South Africa, being nearly as extensive in area as Scotland (*cf.* Sill, Laccolith, Batholith).'

Himus, 1954. Similar: not in *O.E.D.*

Lōrā (Indo-Pakistan: Baluchi)

A hill torrent carrying rain water.

Spate, 1954. 'The bigger wadis or loras' (Baluchistan) (p. 425).

Louderback (W. M. Davis, 1930)

Davis, W. M., 1930. The Peacock Range, Arizona, *Geol. Soc. Amer. Bull.*, 41.

Thornbury, 1954. 'The term louderback was proposed by Davis for displaced segments of a lava flow on two sides of a fault. If it can be established that the lava flow is of late geologic age, there is justification for assuming that associated scarps were produced by faulting.' (p. 258)

Comment. Named after the geologist G. D. Louderback who first described the phenomenon. An American term not in general use. (G.T.W.)

Lough (Irish)

O.E.D. A lake or arm of the sea; equivalent to the Scottish loch (which is the native

Figure 9a: A sample page from L D Stamp (ed) 'A glossary of geographical terms'

fast ice I. which covers the frozen surface of the sea, but remains f. to the actual coast.

fathom A nautical measurement of depth = 6 ft. = 1·829 m.; 100 fthms. = 1 cable; 10 cables = 1 nautical mile.

fault A surface of fracture or rupture of strata, involving permanent dislocation and displacement within the earth's crust, as a result of the accumulation of strain; see ELASTIC REBOUND. Hence *faulting*. See also NORMAL, REVERSED, TEAR and THRUST F., HADE, HEAVE, THROW. [*f*]

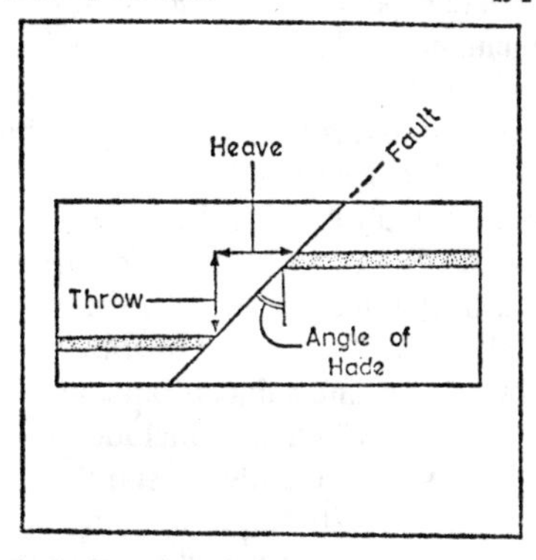

fault-block A section of the earth's crust sharply defined by faults; it may stand up prominently and loftily (BLOCK MOUNTAIN, HORST), be tilted (TILT-BLOCK), or be denuded to the level of the surrounding country (faulted inlier). Ct. GRABEN.

fault-breccia The rock occurring in the SHATTER-BELT of a fault, consisting of crushed angular fragments. [*f* HANGING WALL]

fault-line scarp Where faulting brings rocks of varying resistance into close juxtaposition, differential denudation may wear away the less resistant rock on one side of the fault, thus forming a cliff or scarp along its line. Ct. FAULT-SCARP. The resultant higher ground may be on either the upthrow or downthrow side. E.g. a f.-l.s. has developed along the line of the Mid-Craven F. in Yorkshire. To the N. is a plateau of Carboniferous Limestone, which has been down-faulted to the S., so that less resistant Bowland Shales have been brought into juxtaposition along the line of the f. Denudation has caused the line of the f. to stand out as a f.-l. s., in places 300 ft. high, including Malham Cove and a line of 'scars'. The E. face of the Sierra Nevada, California, is a major f.-l. s., as is the E. face of the Grand Tetons, Wyoming. As denudation progresses, the f.-l. s. may develop into an OBSEQUENT F.-L. S. (facing the opposite direction), then into a RESEQUENT F.-L. S. (facing the original direction). [*f*]

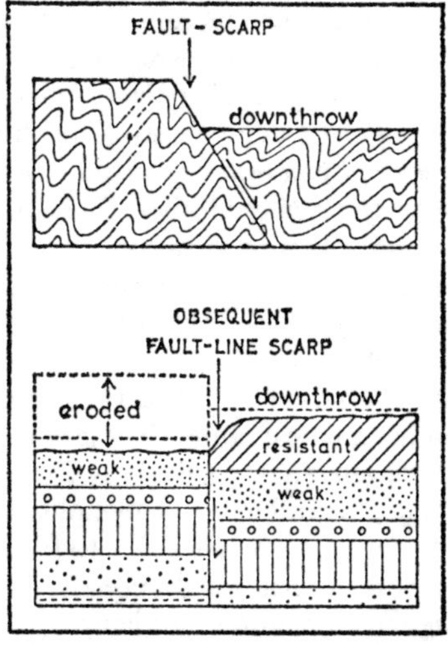

fault-scarp A visible steep edge or slope of a recent fault, due to earth-movements. It is an initial land-form, present only in the earliest stages of denudation succeeding the earth-movement which caused it; e.g. two f.-s.'s were produced in Montana and Wyoming by the Madison earthquake of 1959, forming long 'steps' across country about 4–6 m. high. Most f.-s.'s are soon obliterated by weathering and erosion, or they later develop into FAULT-LINE SCARPS. [*f*]

fault-spring A SPRING thrown out along a fault, where a permeable bed,

Figure 9b: A sample page from F J Monkhouse's 'A dictionary of geography'

to the main alphabetical sequence. *A dictionary of basic geography*, A A Schmieder et al (Allyn and Bacon, 1970), attempts 'to reflect the prevailing thought of American geographers'. No authorities are cited in support of the definitions, but as in the Longman's dictionary a short bibliography (annotated) of basic texts is provided; an advantage of this list by comparison with that in Stamp's dictionary is that the works listed are related by an index to the main sequence of terms, facilitating their use in amplifying the dictionary's explanations. An inexpensive, pocket-sized dictionary which gives very brief definitions is *A concise glossary of geographical terms*, J C Swayne (third edition, G Philip, 1968). No authorities are cited and the glossary is not illustrated. More recent dictionaries are *Dictionary of environmental terms*, A Gilpin (Routledge, 1976), which provides rather patchy coverage, and *An illustrated dictionary of geography*, R O Buchanan (Heinemann, 1974), which is a simple and colourful guide suitable at an introductory level.

C B M Lock's *Geography and cartography: a reference handbook* (third edition, Bingley; Hamden (Conn), Linnet, 1976) is something of a misfit among these dictionaries. This third edition is a fusion of the same author's *Geography: a reference handbook* (second edition, 1972) and *Modern maps and atlases* (1969). The whole has been updated, and like the former volume consists of a series of alphabetically arranged entries for geographers, societies, institutions, publications, and subjects, and an index is also provided. Its scope has now been extended to include cartographers and cartography. Some of the entries are quite substantial, and it is an encyclopedia rather than a dictionary, its emphasis being bibliographical rather than terminological. It is a most useful source for quick reference, particularly for biographical information, which is something of a problem in geography, and the numerous references are a helpful and authoritative guide to further reading.

Several linguistic dictionaries of geographical terms are available, covering the more important languages encountered in the literature. *Glossary of geographical and topographical terms* by A Knox was originally publised in 1904, and has recently been reprinted by Gale Research Company, 1968. This is a single alphabetical list of terms from numerous languages, with very concise, often one word, English equivalents. Many of the terms thus defined are of African and Asian origin and are more frequently met with in the literature of travel and exploration than that of modern scientific geography. German language terminology is conveniently covered in *A German and English glossary of geographical terms*, by E Fischer and F E Elliott (American Geographical Society, 1950); and for Russian, which is an increasingly important language for geographical communication, there is the *Russian-English dictionary of geographical terms*, A Sarna (Telberg Book Corporation, 1965). In this work the Russian words are not transliterated, and there is no English-Russian list. A French work, rather more limited in scope, is *Le langage de géographes: terms, signes, couleurs des cartes anciennes, 1500-1800* by F de Dainville and F Grivot (Picard, 1964). Explanations and definitions are all in French.

An increasingly frustrating feature of modern life is the tendency to use abbreviations and acronyms rather than full names or titles. Fortunately there are a number of standard dictionaries which enable them to be deciphered. Among these are E Pugh's three volume *Dictionary of acronyms and abbreviations* (Bingley; Hamden (Conn), Linnet, 1976); *World guide to abbreviations of organisations* by F A Buttress (fifth edition, Leonard Hill, 1974); *Cassell's dictionary of abbreviations*, J W Gurnett and C E J Kyte (second edition, Cassell, 1972); *Acronyms and abbreviations dictionary* (third edition, Gale Research, 1970, and later supplements); and the two volume *World guide to abbreviations of associations and institutions* (Bowker, 1972).

Gazetteers and place names

The best all-round gazetteer for most purposes is *The Columbia Lippincott gazetteer of the world,* edited by L E Seltzer and produced by Columbia University Press with the cooperation of the American Geographical Society. There are some 130,000 entries, each giving the location (often in terms of the distance from the nearest large town) and brief historical and economic notes. Rivers, mountain ranges and other physical features are included in addition to cities, towns, and other political and administrative areas. The *Times index-gazetteer of the world*, although containing more entries than the *Columbia Lippincott gazetteer* is less helpful in respect of the amount of descriptive and factual information which it incorporates. Unlike the American volume it is merely a location tool, citing coordinates and where appropriate, map references to the *Times atlas.*

Where neither of the above major gazetteers is available, there are several less comprehensive substitutes. *Chambers's world gazetteer and geographical dictionary* (revised edition, 1965) contains about 12,000 entries; *Webster's new geographical dictionary* (revised edition, Merriam, 1976) has rather more and is more up-to-date, but the amount of detail given is less. The *Statesman's yearbook: world gazetteer* compiled by J Paxton (Macmillan, 1975) as a supplement to the well known *Statesman's yearbook* contains not only a select list of place names but an atlas section, miscellaneous statistical tables, and a list of geographical terms. A more specialised gazetteer which is a standard source for the UK is *Bartholomew's gazetteer of Britain* compiled by O Mason (Bartholomew,

1977). This is a new edition of a well-established reference book containing over 40,000 place names, each entry giving a map reference.

The standardisation of place names in various languages is a major problem, and several sources are available to provide solutions. By far the most comprehensive attempt is the series of gazetteers produced by the US Board on geographic names, published in Washington from 1955. Over one hundred volumes have appeared in this series covering various countries and regions of the world. Lists of the volumes available are provided in both Walford and Winchell. A more modest series, covering a smaller number of countries is published by the Royal Geographical Society Permanent Committee on Geographical Names for British Official Use. These list alphabetically the officially approved forms, giving the location of the places cited and noting alternative forms. Not intended as a means of standardisation so much as a guide to identifying variant forms of place names in different languages is *Glossary of geographical names in six languages*, G Lana et al (Elsevier, 1967). The languages included are English, French, Italian, Spanish, German, and Dutch. Arrangement is alphabetical by standard English usage, with the equivalent names in the other languages entered below, an index giving access by way of these alternatives. The selection of names for this book is somewhat curious, and a number are cited in their English forms without any variations.

The rendering of geographical names by M Aurousseau (new edition, Greenwood Press, 1976) is not a reference book in the sense of the works discussed previously, but it provides an interesting and useful discussion of the problem of nomenclature, incorporating numerous references to both further reading on the subject and authoritative lists of standard forms and transliteration systems. Geographical names are in a continual state of evolution, reflecting political change, and new names are created for new communities; both factors lead to problems for the geographer in keeping information in the gazetteers up to date. An attempt to solve these problems is *New place names of the world* by H Spaull (Ward Lock, 1970), which lists new names and name changes since 1945. *Names on the globe* by G R Stewart (Oxford, 1975) is a fascinating account of the origins and meanings of a wide variety of place names.

Directories and yearbooks

Geographisches Taschenbuch und Jahrweiser fur Landeskunde has been published fairly regularly since 1949, annually at first and more recently at longer intervals. The latest edition is for 1975/6. It contains short articles and papers, statistical and biographical information, specialised directories, and expedition and research reports. Harris provides an analysis of principal bibliographies included up to 1972.

Of a much broader scope, but originally issued as a supplement to *Geographisches Taschenbuch, Orbis Geographicus* (published by Franz Steiner Verlag) is an important international geographical directory. Compiled and edited by E Meynen of the Institut fur Landeskunde, Bad Godesberg, for the International Geographical Union, the current edition (1968/1972) is in two volumes. Part one lists societies, academic departments, research institutes, cartographic and topographic surveys, hydrographic and oceanographic offices, important map collections, and national commissions and authorities on geographical names, as well as a considerable amount of information about the IGU itself. The entries for these include the names of senior staff together with their respective specialisations and responsibilities, and the second volume consists of a list of professional geographers arranged by countries.

Many national geographical associations and societies also publish lists and directories of their members. The Association of American Geographers' *Directory* now appears at approximately three yearly intervals, sometimes as supplements to the *Professional geographer* but more recently as a separate publication. The latest edition (1974) lists some 6200 names, with an indication of specific subject interests. The *Transactions of the Institute of British Geographers* formerly included occasional lists of members (a separate list is also available) as does the *Association de Géographes Français bulletin*. A whole issue of *Soviet geography* (volume 8, number 7, 1967) is devoted to a *Directory of Soviet geographers.*

Many non-geographical directories are also relevant. Information on international associations can be obtained from the *Yearbook of international organizations: the encyclopaedic dictionary of international organizations, their officers, their abbreviations* (Union of International Associations, biennial). This is supplemented by updated information in a monthly journal published by the UIA entitled *International associations.* National associations are listed in a variety of publications, such as *Directory of British associations* (fifth edition, CBD Research, 1977) and its companion work the *Directory of European associations.* The latter is in two parts, on *National industrial, trade, and professional associations* (second edition, CBD Research, 1976), and on *National learned scientific, and technical societies* (CBD Research, 1975). For the United States there is the *Encyclopaedia of associations* (Gale Research), which

CONTENTS

Figure 10: Contents page from 'Geographical digest'

is revised about every two years and includes a loose-leaf updating service, and *Research centers directory* (biennial, Gale Research, 1960-) which is mainly a guide to academic institutions and foundations. Directories for many other countries are listed in the standard guides such as Walford and Winchell.

The *World of learning* (Europa, annual), already referred to in another context, is a major international guide to educational, scientific, and cultural institutions. Arranged by country, the various entries include the names of senior staff and their specialities. The *Commonwealth universities yearbook* (Association of Commonwealth universities, annual) is a useful register of academic staff in a number of countries, and incorporates an index to their names. Similar information for the United States is available in *American universities and colleges*, published by the American Council on Education and revised about every four years.

More general in character than any of the above directories are several very widely available yearbooks, including *Whitaker's almanack, The world almanac and book of facts,* and the *Statesman's yearbook.* These compact annual volumes contain a wealth of factual and directory-type information, and generally speaking are well arranged and indexed for rapid reference. A rather more substantial work, again widely available in reference libraries, is the *Europa yearbook.* This is published annually in two volumes, entitled respectively 'International Organisations, Europe', and 'Africa, the Americas, Asia, Australasia'. Following the initial descriptions of international organisations, there is a country by country arrangement, each section including statistical, political, legal, economic, educational, and cultural information.

Within geography itself there is a well-established and very handy yearbook similar in type to the general ones discussed above. This is the *Geographical digest*, edited by H Fullard and published by G Philip annually since 1963. This aims to provide a selection of useful recent factual data of interest to geographers, reflecting significant changes since the previous year. It is well organised, widely available, and promptly published (in May), although it is very decisively biased towards the needs of British users. The range of material included is illustrated by the contents page of the latest edition reproduced in figure ten.

Four

GENERAL GEOGRAPHY: PERIODICALS

THERE IS NO universally acceptable definition of a periodical. For the purposes of this chapter a periodical is taken to mean a publication appearing serially at more or less regular intervals of less than one year, in a numbered sequence planned to continue indefinitely, and normally containing material by several contributors in each issue. This implies the exclusion of irregularly published series of monographs like the extremely valuable *Lund studies in geography.* Also excluded are a number of annual publications containing primarily bibliographical or statistical material, which are discussed elsewhere. There is confusion in terminology as well as in definition, but the terms 'periodical', 'journal', and 'serial' are used here synonymously.

Journals constitute an extensive and vitally important sector of the literature of geography, as of most other disciplines. They are the principal outlet for the reporting of research, particularly academic research, and for news items of professional interest, and are more generally accessible than, say, theses or reports. Even where research is first recorded in the form of a thesis, it is typical for the author to reproduce in a journal article a summary of the main results or at least of some especially interesting aspect of the work. Although the journals discussed in this chapter form only a small proportion of the total containing items of geographical interest, and references to articles in journals, are, as a rule, better gained by the use of abstracting and indexing services than by a direct approach to the journals themselves, it is nevertheless as well to be acquainted individually with the main titles in the field. In the first place any bibliographic listing of the content of journals is subject to publishing delays and cannot includes articles from the most recent issues. And secondly, although articles constitute the largest portion of most journals, other information is also incorporated which is often never indexed by these services, usually because it is of current value and has less lasting significance. The researcher and the professional geographer, therefore, have no alternative but to examine the main journals as they are published, if they wish to derive maximum benefit from the contents and keep fully abreast of new developments in the subject. An impression of the size of this problem may be gained from the results of a survey by W O Aiyepeku of articles indexed in *Current geographical publications* over two years (*Libri*, vol 22, 1972, 169-182). This produced a list of ninety-one titles containing twenty or more significant papers; a repeat of this exercise on more recently published volumes would certainly result in a higher figure.

Content

Libraries differ in their treatment of journals; in some collections they are classified (like books) and in others they are arranged alphabetically by title or in very broad subject groups or in some other way. The problem, obviously, is that the contents of even a single issue of one journal may range widely over a variety of topics. For the geographer, this problem is intensified by the all-embracing nature of his subject, and geographical articles appear in many non-geographical journals. Chauncy Harris estimated that only one quarter of the nine hundred serials cited in the *Bibliographie geographique internationale* were primarily geographical (*Geographical review* vol 58, 1968, 147). Items of interest appear in both specialised scholarly journals in fields other than geography, and in certain popular periodicals of a more general nature such as *New society, New scientist, Ecologist* and *Scientific American.*

It is interesting to note that there has been little development of major journals devoted to systematic geographical studies. Although geography is a subject composed of a number of disparate subdisciplines, the core periodical literature remains general rather than specialist in character. Of course there are notable exceptions (*Economic geography* is perhaps the most striking, although its contents are broader than the title suggests) and many journals do incorporate a bias indicative of editorial policy. A recent trend, which perhaps reflects an increasingly strong tendency for the various specialisms within geography to move further apart, is a considerable extension of these specialised journals: *Journal of historical geography, Journal of biogeography*, and *Earth surface processes: a journal of geomorphology* are three important examples which all began publication within the last three years. In spite of this, the fact remains that such sources are not typical, and extensive use must be made of non-geographical journals for advanced work in any branch of systematic geography, and certainly at research level.

In regional geography, by contrast, there is a considerable development of specialist journal literature. This is largely because most geographical serials are published by societies and institutions with a local base and consequently reflect to some extent local interest. The place of publication is frequently indicative of a regional bias in the contents of journals in geography. This may be on a national scale (*eg Australian geographer; Irish geography*) or confined to a more localised area (eg *East Midland geographer*). In the UK this process can be further illustrated by the growing practice of student geographical societies based on university and college departments of geography to publish journals, usually annually. These, although of varying quality, in many cases make a worthwhile contribution to the literature on what might otherwise be much neglected areas.

The main advantage of the journal over the book as a form of communication is its immediacy: it allows new information to be published and made accessible with a minimum of delay. This being so, there is little point in using a journal to reproduce old material, and a former editor of the *Geographical review* has written: '. . ."Is it new?" is the first question asked about articles submitted for publications. "New", of course, has a wide connotation–it applies to timeliness (news), to content, to interpretation, method, lines of thought . . .' (*Geographical review*, vol 38, 1948, 177). Whether or not they conform to this ideal, many articles have more than a merely topical significance, and most self-respecting journals provide indexes or at least cumulated contents lists to facilitate retrospective searching. A further instance of this longer-term role of periodicals is the development and increasing importance of review series: these contain articles which set out to review and consolidate existing knowledge rather than present original material. Such articles typically incorporate lengthy bibliographies of original sources. Already well-established for many years in most scientific and technological subjects, and usually published on an annual basis, review journals are rapidly gaining ground in other areas as well; the annual *Progress in geography*, which began in 1969, has now proved the worth of this form of publication in geography, and increased its size by splitting into two parallel series *Progress in human geography*, and *Progress in physical geography* (1977). The summaries of key literature and the bibliographies in these greatly assist in retrospective searching.

This function is duplicated by the independent abstracting and indexing services, but whereas these confine themselves in the main to articles, a journal's own index is normally more comprehensive. In particular, it is quite common for book reviews to be included. The majority of the major journals carry such reviews, which constitute an essential ingredient in the raison d'être of periodical literature. The review is the first public impartial and critical assessment of new books, and can be a most helpful guide in selecting the most reliable and useful work where several alternatives are available. It is not difficult to trace reviews of important books by checking in the indexes of the most appropriate journals; and finding out which are the most appropriate journals, although it is largely a matter of educated guesswork, does not pose too many problems in practice. Normally it is not worth checking more than, say, eighteen months to two years from the date of publication of the original work, although the frequency of the journal being checked, the number of reviews carried, and any discrepancy between its official and actual date of publication, should be taken into consideration.

Articles and book reviews are the meat of most journals. Another common feature, stemming from the role of so many journals as association publications, is news. This includes announcements of forthcoming conferences, symposia, and meetings, and of professional appointments, resignations, and retirements. Obituary notices are also useful, especially when they describe a geographer's work and evaluate the significance of his contribution to the subject. Sometimes obituaries include full lists of publications, as for example is the practice of the *Annals of the Association of American Geographers* and the *Transactions of the Institute of British Geographers.*

There is a lot of bibliographical information relating to new publications, new editions, and reprints to be gleaned from journals, apart from the contents of the book reviews section. Few journals can review all the books sent them for this purpose, and many publish brief details acknowledging the receipt of those not in fact reviewed. In addition there are publishers' advertisements and announcements, and the journals of societies which provide library facilities for members sometimes incorporate accessions lists. The earliest pre-publication indication of future work is in lists of research projects in progress, which are published in a number of journals. These lists enable the researcher to make contact with others working in similar fields before any results actually appear in the literature, thus minimising any duplication of effort.

Bibliographies of periodicals

The descriptive list of core journals which forms the latter part of this chapter is highly selective, as are the references to more specialised

ETUDES MALIENNES. see *HISTORY — History Of The Near East*

ETUDES RURALES; revue trimestrielle d'histoire, geographie, sociologie et economie des campagnes see *SOCIOLOGY*

EUROPE SUD-EST. see *POLITICAL SCIENCE*

910 US ISSN 0014-5025
EXPLORERS JOURNAL. 1921. q. membership(non-members $6) Explorers Club, 46 E. 70th St., New York, NY 10021. Ed. Henry S. Evans. adv. bk. rev. charts. illus. index. maps. circ. 2,200. Indexed: Biol.Abstr.

910 FI ISSN 0015-0010
FENNIA. (Text mainly in English; occasionally in German or French) 1889. ca. 10/yr. price varies per no. Academy of Finland, Geographical Society of Finland, Helsinki University, Dept. of Geography, Hallituskatu 11-13, SF-00100 Helsinki 10, Finland. Ed. Paul Fogelberg. charts. illus. index per vol. circ. 1,000(approx.) Indexed: Biol.Abstr. Geo.Abstr. Chem.Abstr.
Incorporating: Acta Georgaphica.

910
FINISTERRA; revista Portuguesa de geografia. (Text in Portuguese; summaries in English, French and Italian) 1966. s-a. Esc 90. Centro de Estudos Geograficos, Faculdade de Letras de Lisboa, Livraria Portugal, Rua do Carmo 70, Lisbon, Portugal. bk. rev. bibl. charts. illus. stat. index. circ. 1,400. Indexed: Geol.Abstr.

FLAMENCO; boletin de informacion. see *FOLKLORE*

917.59
FLORIDA GEOGRAPHER. 1971. q. $0.50 for members; $1.50 for non-members. ‡ Florida Society of Geographers, 2014 N.W. 11th Road, Gainesville, FL 32601. Ed. Louis Paganini. adv. bk. rev. circ. 300.

917.4
FLYING YOUR WAY. 1973. Air New England, Inc., Logan International Airport, East Boston, MA 02128. adv. circ. 10,000.
New england people and places

910 US ISSN 0015-5004
FOCUS (NEW YORK, 1950) 1950. bi-m. $700. American Geographical Society, Broadway at 156th St., New York, NY 10032. Ed. Alice Taylor. bibl. illus. maps. circ. 15,000. (microform; also avail. in microform from XUM) Indexed: P.A.I.S. R.G.

910 HU ISSN 0015-5403
FOLDRAJZI ERTESITO. (Summaries in English, French, German and Russian) 1952. q. $11.80. (Magyar Tudomanyos Akademia, Foldrajztudomanyi Kutato Intezet) Akademiai Kiado, Publishing House of the Hungarian Academy of Sciences, Box 24, H-1363 Budapest, Hungary. Ed. S. Marosi. bk. rev. abstr. bibl. charts. illus. stat. index. (tabloid format) Indexed: Chem.Abstr.

910 HU ISSN 0015-5411
FOLDRAJZI KOZLEMENYEK. (Text in Hungarian; summaries in English, French and German) 1873. q. $11.80. (Magyar Foldrajzi Tarsasag) Akademiai Kiado, Publishing House of the Hungarian Academy of Sciences, Box 24, H-1363 Budapest, Hungary. Eds. M. Pecsi & Gy. Miklos. bk. rev. abstr. bibl. charts. illus. stat. index.

910
FREIBURGER GEOGRAPHISCHE MITTEILUNGEN. 1872. 2/yr. DM.9. Uferstrasse 22, 707 Schwaebisch-Gmuend, W. Germany (B.R.D.) Ed. Jurgen C. Tesdorpf. charts. illus.

GEO ABSTRACTS A (LANDFORMS & THE QUATERNARY) see *ABSTRACTING AND INDEXING SERVICES*

GEO ABSTRACTS B (CLIMATOLOGY AND HYDROLOGY) see *ABSTRACTING AND INDEXING SERVICES*

GEO ABSTRACTS C (ECONOMIC GEOGRAPHY) see *ABSTRACTING AND INDEXING SERVICES*

GEO ABSTRACTS D(SOCIAL & HISTORICAL GEOGRAPHY) see *ABSTRACTING AND INDEXING SERVICES*

526 CS ISSN 0016-7096
GEODETICKY A KARTOGRAFICKY OBZOR. (Text in Czech or Slovak; summaries in English, French, German and Russian) 1913. m. 48 Kcs.($7.) Czech Office for Geodesy and Cartography, Hybernska 2, 111 21 Prague, Czechoslovakia. (Co-Sponsor: Slovak Office for Geodesy and Cartography) Ed. Karel Dvorak. adv. bk. rev. abstr. bibl. charts. illus. stat. index. circ. 2,200.
Cartography

GEODEZIIA I KARTOGRAFIIA. (Glavnoe Upravlenie Geodezii i Kartografii pri Sovete Ministrov SSSR)) see *EARTH SCIENCES — Geophysics*

GEODEZJA I KARTOGRAFIA. see *EARTH SCIENCES — Geophysics*

910.02
GEOGRAFICKE PRACE. (Text in Slovak summaries in English, German, Russian) vol.2,1970. s-a. exchange basis. Kabinet Pre Vyskum Krajiny, Leninovo Nam. 6, Presov, Czechoslovakia. Ed. Jan Sisak, Csc. bibl. charts. illus.

910 CS ISSN 0016-7193
GEOGRAFICKY CASOPIS/GEOGRAPHICAL REVIEW. (Text in Slovak; summaries in English, French, German and Russian) 1949. q. 40 Kcs.($4.28) Slovenska Akademia Vied, Klemensova 27, Bratislava, Czechoslovakia. Eds. E. Mazur & Jozef Kvitkovic. bk. rev. charts. illus. index. circ. 1,000.

910 370 UR ISSN 0016-7207
GEOGRAFIJA V SHKOLE. 1934. bi-m. 2 Rub. 70 kop. Ministerstvo Prosveshcheniya S.S.S.R. Moscow, U.S.S.R. Ed. N.A. Maksimov. bk. rev. bibl. charts. illus. index. circ. 150,000.

GEOGRAFISCH TIJDSCHRIFT. see *PHOTOGRAPHY*

910 DK ISSN 0016-7223
GEOGRAFISK TIDSSKRIFT. (Text in Danish, English and German) 1877. s-a. Kr.100($10) Kongelige Danske Geografiske Selskab - Royal Danish Geographical Society, Haraldsgade 68, 2100 Copenhagen 0, Denmark. Prof. Dr. N. Kingo Jacobsen. bk. rev. abstr. bibl. charts. illus. index. cum.index. Indexed: Geo.Abstr.

910 SW ISSN 0016-7231
GEOGRAFISKA ANNALER. (Series A: Physical Geography; Series B: Human Geography) (Text in English, French and German) n.S.1965. s-a. Kr.100($55) (series A); Kr.40($8) (series B) (Svenska Saellskapet Foer Antropologi och Geografi) Generalstabens Litografiska Anstalt, Box 22069, S-104 22 Stockholm, Sweden. Ed. Dr. John O. Norrman (Series A); Dr. Staffan Helmfrid. (Series B) circ. 800. Indexed: SSCI.

907 370 SW ISSN 0016-724X
GEOGRAFISKA NOTISER; medlemsblad for geografilararnas riksforening. 1943. 4/yr. Kr.20. c/o Nils Lewan, Ed., Dept. of Geography, 223 62 Lund, Sweden. adv. bk. rev. bibl. charts.
For high school and university teachers

910 YU ISSN 0016-7266
GEOGRAFSKI HORIZONTI. 1955. q. 10 din. Geografsko Drustvo Hrvatske, Marulicev Trg 19/2, Zagreb, Yugoslavia. bk. rev.

910 YU ISSN 0016-7274
GEOGRAFSKI OBZORNIK; casopis za geografsko vzgojo in izobrazbo. (Text in Slovenian) vol.13,1966. q. 600 din. Geografsko Drustvo Slovenije, Askerceva 12, Ljubljana, Yugoslavia. Ed. Mara Radinja.

910 AT ISSN 0046-5666
GEOGRAPHER. 1970. 6/yr. (Feb.-Nov.) Aus.$3. ‡ Carlson Marsh and Associates, Urch Rd., Roleystone, W.A. 6111, Australia. Ed. Colin J. Marsh. adv.

910 PL ISSN 0016-7282
GEOGRAPHIA POLONICA. (Text in English and French) 1964. 2-4/yr. price varies. (Polska Akademia Nauk, Instytut Geografii) Panstwowe Wydawnictwo Naukowe, Miodowa 10, Warsaw, Poland. Ed. Stanislav Leszczcki. charts. illus. circ. 1,000. Indexed: Chem.Abstr. Geo.Abstr.

910 NR ISSN 0016-7290
GEOGRAPHICA. 1967. q. University of Ife, Geographical Society, Ife, Nigeria. Ed. Lanre Filani. adv. bk. rev.

910 SZ ISSN 0016-7312
GEOGRAPHICA HELVETICA; Schweizerische Zeitschrift fuer Laender- und Voelkerkunde. (Text and title in French, German and Italian) 1946. q. 40 Fr. (Geographisch-Ethnographische Gesellschaft Zurich) Kuemmerly und Frey AG, Hallestr. 6-10, CH-3001 Bern, Switzerland. adv. bk. rev. abstr. bibl. charts. illus. stat. index. circ. 1,500. Indexed: Biol.Abstr.

GEOGRAPHICAL ABSTRACTS; geography, remote sensing, and cartography. see *ABSTRACTING AND INDEXING SERVICES*

910 US ISSN 0016-7363
GEOGRAPHICAL ANALYSIS; an international journal of theoretical geography. 1969. q. institutions & libraries $14; individuals $12. Ohio State University Press, 2070 Neil Ave., Columbus, OH 43210. Ed. R. Golledge. adv. bk. rev. index. circ. 1,250. (back issues avail) Indexed: Geo.Abstr. SSCI.

910 TZ ISSN 0016-738X
GEOGRAPHICAL ASSOCIATION OF TANZANIA JOURNAL. 1967. s-a. membership. Geographical Assn. of Tanzania, University of Dar es Salaam, Box 35049, Dar es Salaam, Tanzania. Eds. W. Mlay P. Maro. bk. rev. charts. illus. stat. circ. 400. (processed)

910 UK ISSN 0016-7398
GEOGRAPHICAL JOURNAL. 1893. 3/yr. £10($27) Royal Geographical Society, 1 Kensington Gore, London SW7 2AR, Eng. Ed. Sir Laurence Kirwan. adv. bk. rev. charts. illus. index. cum.ind. every 10 years. circ. 10,000. Indexed: Br.Hum.Ind. P.A.I.S. Soc.Sci.Ind. SSCI.

910 II ISSN 0016-7401
GEOGRAPHICAL KNOWLEDGE. (Text in English & Hindi) 1968. s-a. Rs.15($6) Society for Geographical Studies, Kanpur, 7-125 Swarup Nagar, Kanpur-2, India. Ed. Dr. D. P. Saxena. adv. bk. rev. charts. stat.

910 UK ISSN 0016-741X
GEOGRAPHICAL MAGAZINE. 1935. m. L.4.50($12) (Royal Geographical Society) New Science Publications, 128 Long Acre, London WC2E 9QH, Eng. Ed. Derek Weber. adv. bk. rev. illus. maps. index. circ. 76,376. (also avail. in microfilm from WMP) Indexed: Br.Hum.Ind. P.A.I.S. Soc.Sci.Ind.

910
GEOGRAPHICAL PAPERS. 1970. 10-12/yr. University of Reading, Department of Geography, Whiteknights, Reading, Eng. adv. Indexed: Geo.Abstr.

910
GEOGRAPHICAL PERSPECTIVES. 1963. s-a. $4. Univ. of Northern Iowa, Dept. of Geography, Dept. of Geography, Sabin 1, Cedar Falls, IA 50613. (Co-Sponsor: Iowa Council for Geographic Education) Ed. Basheer K. Nijim. circ. 200.
Formerly (until 1974): Iowa Geographer (ISSN 0047-1399)

910 US ISSN 0016-7428
GEOGRAPHICAL REVIEW. 1916. q. 17.50. American Geographical Society, Broadway and 156th St., New York, NY 10032. Ed. Sarah K. Myers. bk. rev. illus. maps. index. cum.ind. every 10 years; v.1-55(1916-1965)(in 5 vols) circ. 7,200. Indexed: Abstr.Anthropol. Amer.Hist. & Life. Biol.Abstr. Bk.Rev.Ind. Chem.Abstr. Curr.Cont. Eng.Ind. Hist.Abstr. P.A.I.S. Peace Res.Abstr. Soc. Sci. Ind. SSCI.

Figure 11a: Sample page from 'Ulrich's periodical directory'

GANN MONOGRAPHS ON CANCER RESEARCH. XXX
+ + GANN MONOGR CANCER RES
NIHON GAN GAKKAI
TOKYO NO 11 1972-
PREV GANN MONOGRAPHS FROM 1 1966- 10. 1971
CONSISTS OF PROCEEDINGS OF INTERNATIONAL CONFERENCES & SYMPOSIA
CA/U-1 LO/M-3

GARAGE MANAGEMENT. XXX
+ + GARAGE MANAGE
[H I THOMPSON P]
LONDON 1(1). S 1974
MON PREV ACCESSORY & GARAGE EQUIPMENT FROM 1. JA 1956- 19. AG 1974 S/T ACCESSORIES SALES. SERVICING. REPAIRS. EQUIPMENT
LO/N14
CA/U-1 2(2). 1975- ED/N-1 2(2). 1975-
ISSN 0306-0268

GARAGE & TRANSPORT. XXX
+ + GARAGE & TRANSP
[HULTON]
SWANLEY KENT 20(3) 1974-
MON PREV GARAGE & TRANSPORT EQUIPMENT FROM 1(1). 1955 20(2). 1974
LO N14

GARAGE & TRANSPORT EQUIPMENT. XXX
SUBS (1974): GARAGE & TRANSPORT.

GARCIA DE ORTA: SERIE DE ANTROPOLOGIA.
+ + GARCIA ORTA ANTROPOL
JUNTA DE INVESTIGACOES DO ULTRAMAR (PORTUGAL)
LISBON 1 1973
LO/N 2

GARCIA DE ORTA SERIE DE ESTUDOS AGRONOMICOS.
+ + GARCIA ORTA ESTUD AGRON
JUNTA DE INVESTIGACOES DO ULTRAMAR (PORTUGAL)
LISBON 1 1973
LO N-2 LO R6

GARDEN JOURNAL OF THE ROYAL HORTICULTURAL SOCIETY. XXX
LONDON 100(6). JE 1975
MON PREV JOURNAL OF THE ROYAL HORTICULTURAL SOCIETY NS FROM 1(1). 1866- 100(5). 1975
ED/N-1 LO/N-2 LO/N13 LO/R6

GARDEN HISTORY. JOURNAL OF THE GARDEN HISTORY SOCIETY. XXX
+ + GARD HIST
LONDON 1 S 1972-
PREV NEWSLETTER. GARDEN HISTORY SOCIETY FROM NO 13. 1970- 17. 1972
CA/U-1 LO/N-2 OX U-8
BL/U-1 3 1974- NO/U-1 2. 1973-

GASTRIC SECRETION.
UNIVERSITY OF SHEFFIELD BIOMEDICAL INFORMATION SERVICE
SHEFFIELD JA 1974
MON COMPRISES MATERIAL EXTRACTED FROM INDEX MEDICUS BIOMEDICAL INFORMATION SERVICE
SUBORD TO DEPARTMENT OF PHYSIOLOGY
CA/U-1

GASTRO-ENTEROLOGIA (ROME). XXX
ITALIAN SOCIETY OF GASTROENTEROLOGY
[EDIZIONI LUIGI POZZI]
ROME 1, 1969
SUBS: RENDICONTI ROMANI DI GASTRO-ENTEROLOGIA

GASTROENTEROLOGIE & STOFFWECHSEL.
+ + GASTROENTEROL & STOFFWECHSEL
[THIEME]
STUTTGART 1. 1972-
LO/N13

GASTROINTESTINAL HORMONES.
+ + GASTROINTEST HORM
UNIVERSITY OF SHEFFIELD BIOMEDICAL INFORMATION SERVICE
SHEFFIELD JA 1973-
MON COMPRISES MATERIAL EXTRACTED FROM INDEX MEDICUS BIOMEDICAL INFORMATION SERVICE
SUBORD TO DEPARTMENT OF PHYSIOLOGY
CA/U 1 1974

GEC JOURNAL OF SCIENCE & TECHNOLOGY (1972). XXX
+ + GEC J SCI & TECHNOL (1972)
GENERAL ELECTRIC COMPANY
LONDON 39(2) 1972-
PREV JOURNAL OF SCIENCE & TECHNOLOGY FROM 36(1). 1969- 39(1) 1972
GL/U-1 LO/N14
ISSN 0302-2587

GEFLUGELHOF. XXX
SUBS (1969): GEFLUGELHOF UND KLEINVIEH.

GEFLUGELHOF UND KLEINVIEH. XXX
SCHWEIZERISCHER GEFLUGELZUCHTVERBAND
BERNE 32. 1969- 36. 1973
PREV GEFLUGELHOF FROM 1. 1937- 31. 1968 SUBS SCHWEIZERISCHE GEFLUGEL-ZEITUNG
LO/N13
ISSN 0016-5832

GENERAL PHARMACOLOGY. XXX
+ + GEN PHARMACOL
[PERGAMON]
OXFORD &C. 6(1). MR 1975-
Q PREV COMPARATIVE & GENERAL PHARMACOLOGY FROM 1. MR 1970- 5(3/4). D 1974
ED/N-1 LO/N14 LO/U-1
ISSN 0306-3623

GENERAL PRACTITIONER. XXX
+ + GEN PRACT
[HAYMARKET PUBL]
LONDON 8/JA 1971-
WKLY PREV GP NEWSPAPER FOR THE GENERAL PRACTITIONER FROM 1(1) N 1963- 51(6). D 1970 S/T THE POSTGRADUATE NEWSPAPER FOR THE G.P
OX/U-8
ED/N-1 20/MR 1971-
ISSN 0046-5607

GEOCOM PROGRAMS.
[GEOSYSTEMS]
LONDON NO 1. 1971-
CA/U-1 OX/U-8
ISSN 0305-0017

GEOGRAPHICAL PAPERS. DEPARTMENT OF GEOGRAPHY. UNIVERSITY OF READING.
+ + GEOGR PAP DEP GEOGR UNIV READING
READING NO 1. 1970-
RE/U-1
CA/U-1 NO 21. 1973-
LO/U-2 NO 21. 1973-
ISSN 0305-5914

GEOLOGIE MEDITERRANEENNE.
+ + GEOL MEDITERR
UNIVERSITE DE PROVENCE
MARSEILLES 1. 1974-
Q S/T ANNALES DE L UNIVERSITE DE PROVENCE
LO/N-2 SW/U-1

GEORGIA JOURNAL OF INTERNATIONAL & COMPARATIVE LAW.
+ + GA J INT & COMP LAW
UNIVERSITY OF GEORGIA: SCHOOL OF LAW
ATHENS, GA 1. 1970-
2/A
CA/U13
ISSN 0046-758X

GERMAN POLITICAL STUDIES.
+ + GER POLIT STUD
GERMAN POLITICAL SCIENCE ASSOCIATION
[SAGE PUBL]
LONDON 1. 1974-
CA/U-1 GL/U-1 HL/U-1 MA/P-1 OX/U-1
ISSN 0307-7233

GERMANO-SLAVICA.
UNIVERSITY OF WATERLOO (ONTARIO) DEPARTMENT OF GERMANIC & SLAVIC LANGUAGES & LITERATURE
WATERLOO, ONT NO 1. 1973-
2/A. ENGL OR GER
LO/U10 SA/U-1

GHANA ECONOMIC REVIEW.
+ + GHANA ECON REV
[EDITORIAL & PUBL SERVICES]
ACCRA 1970-
ANNU.
LO/U-3
LO/N-1 1971/72-
OX/U16 [1970 & 1971/72]

GIS NEWSLETTER. XXX
+ + GIS NEWSL
GEOSCIENCE INFORMATION SOCIETY (US)
COLUMBUS, OHIO NO 27. D 1973-
2M PREV NEWSLETTER. GEOSCIENCE INFORMATION SOCIETY FROM NO 1. MY 1966- 26. OC 1973
LO/N-2 NO 28. 1974-

Figure 11b: Sample page from 'BUCOP' annual supplement. These supplements are known as 'New periodical titles'

periodicals cited elsewhere in the book. These constitute only a small proportion of the total number used by geographers for research purposes. There are, however, a number of bibliographies and guides to periodicals from which more comprehensive information can be obtained. They are, of course, a distinctly different category from those bibliographic tools which index the actual content of specific journal issues; those mentioned here only list periodical titles and publication details, with or without some degree of annotation on content.

Ulrich's international periodicals directory, published by Bowker, is perhaps the best known of these guides, and is revised every two years (latest edition 1977). It is arranged in subject groups, with an alphabetical index. A sample page is shown in figure eleven (a), illustrating the detail provided about each title. The entry for *Geographica helvetica*, for instance, gives ISSN (international standard serial number), title, frequency of publication, price, publisher, and some indication of content–advertisements, book reviews, abstracts, bibliographies, charts, illustrations, and an index. Its circulation is 1500 and it is indexed by *Biological abstracts.*

A companion volume to the periodicals directory is *Irregular serials and annuals: an international directory*, which lists such elusive material as conference proceedings, and yearbooks. Another well-known standard work is the *World list of scientific periodicals published in the years 1900-1960* (fourth edition in three volumes, Butterworths, 1963-5). A fourth volume, *New periodical titles 1960-1968* was added in 1970. An important by-product of this list, incidentally, is in its recommended standardisation of abbreviated titles, which will be found in references from many sources. The list has now been incorporated into another work, the *British union catalogue of periodicals*, also published by Butterworths, often known more simply as *BUCOP*. This originally appeared 1955-8, and now issues quarterly up-dating supplements, with annual and five yearly cumulations. (Butterworth's address in the USA is 19 Cummings Park, Woburn, Mass 01801). The great advantage of *BUCOP* for users in the UK is that it is a list based on the holdings of major British libraries, and locations where the periodicals cited may be consulted are noted in each entry, as may be seen from the sample page shown in figure eleven(b). Following the most obvious bibliographical details, the symbols at the end of each entry represent holding libraries.

A further location list worth noting is that published by the British Library Lending Division, *Current serials received by the BLL*. The latest edition of this was published in 1974, and contains details of nearly

45,000 titles, all obtainable on interlibrary loan from Boston Spa. For North American libraries there is the *Union list of serials in libraries of the US and Canada* (third edition, Wilson, 1965), which incorporates information on over 150,000 journals in nearly 1000 libraries. This is kept up to date by a current publication, *New serial titles*, listing new titles received by the Library of Congress and a variety of other libraries, and available in several regular cumulations. In addition to the above guides, which are all international in scope, there are lists devoted to the journal output of individual countries: for example, the American *Standard periodical directory* (fifth edition, Oxbridge Communications, 1975), and, for the UK, *Guide to current British periodicals*, D Woodworth (second edition, Library Association, 1973).

Bibliographies of specifically geographical periodicals are not so numerous. It is common for abstracting and indexing services to print occasional lists of the journals from which their citations are drawn, and such a list is incorporated in the annual index to *Geographical abstracts.* A rather shorter list, based on similar criteria, was published in the *Professional geographer* (volume 18, number 4, July 1966, 248-51) entitled 'Most cited serials in geography as represented by *Current geographical publications*'. Although one might reasonably expect this to include all the most important journals, the actual number of citations for each title is not necessarily indicative of relative importance. Serials cited eight times or more in 1964 are included. The *Bibliographie géographique internationale* published in 1966 a supplement containing a list of cited journals, entitled *Répertoire des principaux périodiques d'intérêt géographique cité dans la Bibliographie géographique internationale.* This consists of rather more than 1000 entries, in a single alphabetical sequence by title; addresses and languages of publication are given. The Royal Geographical Society published in 1961 a guide to its holdings under the title *Current geographical periodicals: a hand-list of current periodicals in the library of the Royal Geographical Society.* The advent of new periodicals is noted in a rather haphazard way in several of the main society journals (eg *Geographical journal*; *Geography*) and more systematically by the *Geographical review,* which carries an annual feature briefly describing the year's new titles.

The best bibliographies of geographical serials are both American. The most comprehensive list, containing nearly 2500 serials (the definition is somewhat broader than that used in this chapter) from ninety countries, is the *International list of geographical serials* by C D Harris and J D Fellmann (second edition, University of Chicago Department of Geography, 1971).

78 BEITRÄGE zur alpinen Karstforschung. (Bundesministerium für Land- und Forstwirtschaft. Speläologisches Institut). Wien. no1+ (1956+) Irregular. (N65-1:320a). no2 already issued in 1953. Reprint series.
[Hofburg, Bettlerstiege, Wien I.]

79 CARINTHIA II. Mitteilungen des Naturhistorischen Vereines für Kärnten. Klagenfurt. 81+ (1891+) Annual. (U-2:928b; B-1:506a). Formerly, society as Naturhistorisches Landesmuseum für Kärnten. Supersedes Carinthia. Zeitschrift für Vaterlandskunde, Belehrung und Unterhaltung (not in this list).

80 ———.
Sonderhefte. Klagenfurt. no1+ (1930+) Irregular.

81 DEUTSCHE Rundschau für Geographie. Leipzig; Wien. 1-37 (1878-1915). Closed 1915. (U-2:1315b; B-2:37a).
1-32 as Deutsche Rundschau für Geographie und Statistik.

82 DONAURAUM. (Wien. Forschungsinstitut für Fragen des Donauraumes. Zeitschrift). Salzburg. 1+ (1956+) Quarterly. (N60-1:617c; N68-1:669a; BS:266a).
Supersedes institute's Mitteilungsblatt (not in this list).
Includes a separate English-language synopsis.

83 GEOGRAPHISCHER Jahresbericht aus Österreich. Wien. 1+ (1894+) Biennial. (U-2:1696c; B-2:270b).
4-13 include Wien. Universität. Verein der Geographen. Berichte 29-50 (1902-1924) (116).
[Geographisches Institut der Universität, Wien.]

84 GESELLSCHAFT für salzburger Landeskunde. Salzburg. Mitteilungen. 1+ (1861+) Annual. (U-2:1718b; B-2:291a).
Index: each 10th v has index to preceding 10.

85 ———.
———. Beiheft. 1+ (1946+). (U-2:1718b).

86 ———.
———. Ergänzundsband. 1+ (1960+). (N68-1:877c).

87 GLOBUSFREUND. Publikation. (Coronelli Weltbund der Globusfreunde; Coronelli world league of friends of the globe). Wien. 1+ (1952+) Annual. (N65-1:1102b; BS:354a).
In English or German.
[Gusshausstrasse 20, Wien IV.]

88 GRAZ. Universität. Geographisches Institut.
Arbeiten. Linz a. Donau. 1 (1951). Only one issued (N60-1:830b).

89 ———. ———. ———.
Veröffentlichungen. 1-3 (1925-1930). Closed 1930. (U-2:1759b).

90 JAHRBUCH für Landeskunde von Niederösterreich. (Verein für Landeskunde und Heimatschutz von Niederösterreich und Wien). Wien. ns1-33 (1902-1957). (U-3:2141c; B-1:362a).
Succeeds society's Blätter (1865-1901) (110).
Index: 1902-1927, 1928-1938.

91 KARTOGRAPHISCHE Mitteilungen. Wien. v1-2 no2 (1930-1932). Closed 1932. (U-3:2273b).

92 KARTOGRAPHISCHE und schulgeographische Zeitschrift. Wien. 1-10 (1912-1922). Closed 1922. (U-3:2273b).

93 LINZ. Oberösterreichischer Musealverein. Linz.
Jahrbuch. 1+ (1835+) Annual. (U-4:3127b).
Name of society varies: 1835-1839 as Verein des vaterländischen Museums für Österreich ob der Enns und das Herzogthum Salzburg; 1839-1920 as Museum Francisco-Carolinum; 1937-1944 as Verein für

Figure 12a: A sample page from the 'International list of geographical serials'

IV. SERIALS IN GERMAN

Austria

258. *ÖSTERREICHISCHE GEOGRAPHISCHE GESELLSCHAFT (Until 1959: Geographische Gesellschaft in Wien). MITTEILUNGEN. Wien. 1— (1857—). 3 per annum. Schriftleitung: Erik Arnberger. Neues Institutsgebäude, Universität Wien, Universitätsstrasse 7, 1010 Wien, Austria. Orders: Österreichische Geographische Gesellschaft, Karl Schweighofergasse 3, 1070 Wien, Austria. Abstracts in German, English, and French.

    Long established major scholarly geographical periodical. Articles on all parts of the world. Shorter notes and reports. School geography. Geographical information. News. Book reviews.

259. WIENER GEOGRAPHISCHE SCHRIFTEN (Wien. Hochschule für Welthandel. Geographisches Institut). Wien. 1— (1957—). Irregular but usually several each year. Herausgegeben von Leopold Scheidl. Order from: Verlag Ferdinand, Hirt, Wien, Austria. Price of each number varies. Supplementary English titles in table of contents and English abstracts.

    Research monographs mainly on aspects of the economic geography of Austria. Methodologically interesting.

260. GEOGRAPHISCHER JAHRESBERICHT AUS OESTERREICH. Wien. 1— (1897—). Biennial. Herausgegeben von Hans Bobek und Hans Spreitzer. Schriftleitung Hans Fischer. Geographisches Institut der Universität Wien. Wien, Austria.

    Research articles on Austria. Detailed reports on dissertations, scientific publications, and lectures in each of the Austrian universities, Wien, Graz, Innsbruck, Salzburg, and the Hochschule für Welthandel in Wien.

German Federal Republic

261. *ERDE. ZEITSCHRIFT DER GESELLSCHAFT FÜR ERDKUNDE ZU BERLIN. Berlin. 1— (1853—). 4 per annum. Herausgegeben . . . durch Georg Jensch. Schriftleitung: Frido Bader und Dietrich O. Müller. Gesellschaft für Erdkunde zu Berlin, Arno-Holz-Strasse 14, 1 Berlin 41, Germany. Order from: Walter de Gruyter and Company, 1 Berlin 30, Germany. DM. 54 a year. Supplementary English titles in table of contents and English summaries preceding each article.

    A major international scholarly journal of long standing and high current value. Articles. News of scientists. Reviews. Geographical literature. Society proceedings. Each issue contains an interpreted air photograph.

262. *ERDKUNDE: ARCHIV FÜR WISSENSCHAFTLICHE GEOGRAPHIE. Bonn. 1— (1947—). 4 per annum. Index: 1–17 (1947–1963). Herausgegeben von Carl Troll, Helmut Hahn, Wolfgang Kuls, und Wilhelm Lauer: Schriftleitung: Redaktion: Helmut Hahn. Kartographie: Karlheinz Paffen. Geographisches Institut der Universität, Franziskanerstrasse 2, 53 Bonn, German Federal Republic. Order from: Ferd. Dümmlers Verlag, Kaiserstrasse 33-37, 53 Bonn, German Federal Republic. DM. 52 a year. Supplementary English titles in table of contents and extensive English summaries precede articles in German. Some articles in English.

    A leading international scientific periodical. Wide range of interests. Articles. Reports. Notes. Reviews.

59

Figure 12b: A sample page from the 'Annotated world list of selected current geographical serials in English, French and German'

Based on a variety of published sources, supplemented by information provided by authorities within the countries represented, this is a definitive work incorporating both current and closed titles. The scope of the *ILGS* is fairly strictly limited to 'geographical' serials, and the arrangement is alphabetical by country of origin. As will be seen from figure twelve, basically the information presented is only bibliographical, but there are some short annotations.

Brief descriptions of some of the more important serials are provided in another bibliography compiled by Professor Harris, *Annotated world list of selected current geographical serials in English, French, and German* (third edition, Chicago University Department of Geography, 1971). This includes serials in other languages where supplementary use is made of one or more of the major international languages, and is a considerably expanded version of previous editions, which were limited to journals using English. The arrangement is again alphabetical by country within each language group, with an index of titles and issuing bodies. Annotations are mostly very brief, but provide a fair indication of the subject content, scope, and status of each title. The list contains notes on 316 journals, selected from a far larger number of potential candidates. It is not possible here to discuss even this number of periodicals individually, and selection of the most important titles is not easy. Those referred to here are all fairly broad in scope, having more than local interest and spanning the whole field of geography or a considerable part of it. Preference is given to English language material and well-established authoritative journals in other languages.

Leading British and American journals

There are remarkable similarities between the leading British and American geographical journals. The *Geographical journal*, published by the Royal Geographical Society, may be compared with the American Geographical Society's *Geographical review.* Both contain extensive book reviews, both include society news and notes, both are quarterly, and both have a long history of continuous publication. *Geographical journal* dates from 1893, and can be traced back under other titles (*Journal of the RGS, Proceedings of the RGS*) to 1830. In content it reflects this nineteenth century tradition to some extent in its emphasis on exploration and discovery. Variously entitled in its early days the *Transactions, Proceedings, Journal,* and *Bulletin of the American Geographical Society* (from 1852), the *Geographical review* as such dates from 1916.

Further useful comparisons can be made between journals published by other professional geographical associations in the two countries.

The *Annals of the Association of American Geographers* is a quarterly journal dating from 1911. It contains a variety of articles, both long and short, by American geographers but of international interest. It has no book reviews as such but includes important review articles. Presidential addresses, and abstracts of the text of papers read to association meetings are a regular feature. The *Transactions of the Institute of British Geographers* is the corresponding British periodical, beginning publication in 1935. Originally a monograph series (*Publications of the IBG*), this journal has been variously called *Transactions*, and *Transactions and papers*. Frequency of publication has varied but is currently two numbers per year, and occasional monographs have been issued from time to time in addition to the journal proper. Again there are no reviews, and presidential addresses and other papers delivered at meetings are included as well as articles in the usual sense.

Both the IBG and the AAG publish other journals in addition to the above, and these again are similar both in content and format. *Area,* published quarterly by the former body since 1969, contains short articles and 'serves as a medium for the expression of professional opinions on matters of public interest and as a channel for the communication of reports on the activities of members . . .' The *Professional geographer* is similar in content, although unlike *Area* it does include brief book reviews. Originally known as the *Bulletin of the American Society for Geographical Research*, the present title was assumed in 1949, and the journal now appears quarterly.

All the journals discussed above are scholarly publications produced as outlets for professional research and for a professional readership. There are, however, certain popular periodicals which although intended for a wider audience should not be ignored by the serious student. The *Geographical magazine* and the American *National geographic magazine* are both glossy, lavishly-illustrated popular journals; the designation 'magazine', carefully avoided elsewhere in this chapter, is appropriate in these cases. The changing character of both journals makes an interesting comparison. The *National geographic* began publication in 1888, and one of the earliest contributions was W M Davis' classic paper on the erosion cycle. In the early days it was a serious journal containing regular reviews and very different from more recent issues. Its British counterpart is rather more useful. Although it is also designed essentially for mass consumption and it is hardly the place for the publication of original research articles, the papers and reviews it contains are authoritative and often helpful at an introductory level.

Geography, the journal of the Geographical Association, began life as the *Geographical teacher*, and it still places emphasis on articles for teachers and about the teaching of geography. There are also many papers of general interest, as well as reviews, association news and information about recent publications added to the association's library. It regularly features short articles on current developments under the general heading 'This changing world', similar to 'Geographical record' in the *Geographical review.* The *Journal of geography*, although rather different in appearance, is published by the National Council for Geographic Education and reflects the same bias towards geographical teaching. It dates from 1902 (one year later than *Geography*) and is published monthly between September and May.

Other English-language journals

Nearly all of the periodicals discussed in the preceding section are well-established publications with a long tradition of authoritative and scholarly articles. More recent developments in geography, notably the quantitative 'revolution' in methodology, have led to the growth of a new style of journal, concerned principally with theoretical and statistical aspects of the subject, and consciously interdisciplinary in approach. The *Journal of regional science*, published irregularly from 1958 (three issues per annum since 1969) by the Regional Science Research Institute, is perhaps the best example. It contains articles by contributors from various subject fields in addition to geography, and contents lists from other journals in related areas such as economics and planning. *Geographical analysis*, a quarterly journal which began publication in 1969, is more specifically geographical. It includes a valuable section of 'Research notes and comments' as well as articles, book reviews, and other bibliographical information.

Most of the Commonwealth and other English-speaking countries of the world publish notable geographical journals, usually sponsored by national geographical societies or based on university departments. The majority of these, however, tend to specialise to some extent in matters of local interest; such titles as *New Zealand geographer, South African geographical journal*, and *Indian geographical journal* all come into this category, and although they contain some material of international importance their national bias is sufficient to put them beyond the scope of this general survey. The *Canadian geographer* is something of an exception. This is a quarterly journal published by the Canadian Association of Geographers, and as one might expect articles are in either

English or French, with summaries provided in the other language. As well as papers relating to the geography of Canada it features scholarly contributions of more general interest by Canadian geographers. The *Journal of tropical geography* is a further important example of a regional journal with an international reputation; two issues are published each year, from the Department of Geography at the University of Singapore.

Another primarily English-language journal deserving of special mention is the Swedish *Geografiska annaler*, published by the Svenska Sällskapet för Antropologi och Geografi. It began publication in 1919, and traditionally placed an emphasis on physical geography, only one of the quarterly issues each year being devoted to human topics. From 1965 onwards it has split into two sections, series A on physical and series B on human geography. Articles appear in several languages besides English, and the authorship, although predominantly Swedish, is international.

The one important journal which is international in the sense that its publishing body is international is the *IGU bulletin*, formerly known as the *IGU newsletter*. This is the official news organ of the International Geographical Union, published twice yearly in English or French. It contains no articles as such, being devoted exclusively to announcements and reports of international conferences, reports on the activities of the various commissions of the IGU, and news items of professional interest from member countries.

Soviet geography: review and translation, published by the American Geographical Society, is a convenient tool enabling English-speaking geographers to keep in touch with current Russian work. It appears monthly (except July and August), and is designed to make available in English reports of Soviet research in geography, with an emphasis on methodological problems. In addition to such reports and actual translations of key articles, it includes contents tables of various Russian journals and other bibliographic references to new literature.

Major European journals

Many European countries have long traditions of geographical research and scholarship and professional associations with published journals comparable in stature with the English language titles already described. The best known and most respected of the French ones is *Annales de geographie*, founded in 1891 by Vidal de la Blache and numbering among its subsequent directors such famous names as Emmanuel de Martonne, Albert Demangeon, and Max Sorre. It includes articles and reviews on all aspects of geography, and annual lists of theses. There are six issues per year. Most of the other French periodicals are based on local societies and institutions, typifying the strong regional tradition in French geography; good examples are *Revue de géographie de Lyon, Revue géographique des Pyrénées et du sud-ouest*, and *Revue de géographie alpine*. Useful French periodicals of more general interest are *l'Espace géographique* , published quarterly since 1972 and including English-language summaries, and the rather more established *Information géographique.* Published since 1936, there are five issues a year. It is international in scope, although with a French–and French colonial–bias, and among its more useful features is its tabulation of statistical data on various topics.

There are several German periodicals of long standing. *Die Erde* is one of the oldest, dating from 1853, and previously known under its present sub-title *Zeitschrift der Gesellschaft für Erdkunde zu Berlin.* It is published quarterly, and features society news and reviews as well as major articles. The latter are in German, but English abstracts are provided and English titles are included in the contents list. *Petermanns geographische Mitteilungen*, founded in 1855, is another work which commands international respect. Published quarterly by the East German Geographical Society (Geographische Gesellschaft der Deutschen Demokratischen Republik), each article has English and Russian summaries. It is particularly strong on Soviet and East European geography, and contains much valuable statistical and bibliographical information. *Geographische Zeitschrift* has been published since 1895, except for a break between 1945 and 1962. Again English summaries are included. *Erdkunde: Archiv für wissenschaftliche Geographie* is of more recent origin (1947) and some articles are in English. There are also short reviews, mostly of German books, and detailed English summaries of the articles in German.

A number of other countries have professional associations of geographers which publish journals of international repute, often of long-standing. The majority of these, are, in various degrees, most useful for regional research. Among those having a wider significance and therefore worthy of mention here are: *Rivista geograficà Italiana* (Società di Studi Geografici, 1895-); *Società Geografica Italiana. Bollettino* (1868-); *Geografisk tidsskrift* (Kongelige Danske Geografiske Selskab, 1877-); *Société Belge d'Etudes Géographiques. Bullétin* (1931-); *Koninklijk Nederlandsch Aardrijkskundig Genootschap. Tijdschrift* (1876-); *Norsk geografisk tidsskrift* (Norske Geografiske Selskab, 1926-). Such a list could be extended almost indefinitely, since articles on general geography are liable to occur in a very wide range of periodicals, no matter how specialised or localised general editorial policy and interests appear to be. Any important general articles not included or listed in the journals described in this chapter will probably be located through the abstracting and indexing services described in chapter three, or through similar services in appropriate subject fields other than geography, referred to elsewhere in the book.

Newspapers

Before leaving the subject of journals, some mention should be made of the role of newspapers and the availability of these as contemporary accounts of events and developments of interest to geographers. Newspapers are a very special kind of periodical, and in the context of geographical studies have two distinct functions. Firstly, they provide current information about a changing world, which supplements established knowledge; there are no special problems attached to this aspect of their role. But secondly, newspapers provide a permanent first-hand account of events which has a retrospective function in cases where this subsequently becomes the only detailed, day-to-day, record. For this purpose it is necessary to be able to retrieve information from the backfiles. While not forgetting the special place of the local press, often available in local record offices or public libraries, it is worth mentioning the more generally accessible major national dailies, now sometimes stored in microfilm editions. Papers such as the *Times* and the *New York times* are provided with subject indexes which facilitate such retrospective searching, and an alternative source of similar information is the very useful *Keesing's contemporary archives.* This is a weekly record of events of national and international importance, compiled from reports from news agencies, official sources, and newspapers. Sufficient information is given to serve as an acceptable summary and to identify the original sources. Keesing's is provided with an index which cumulates frequently, and facilitates searching through the files back to the beginnings of the service in 1931.

Five

GENERAL GEOGRAPHY: MONOGRAPHS, TEXTBOOKS AND COLLECTIONS

CONTINUING THE THEME of information sources relating to geography as a whole, attention is focused in this chapter on introductory textbooks and monographs which provide concise descriptions and explanations of the essential nature of the discipline, its philosophy, scope and function. Excluded are a number of so-called introductions to geography which are in fact world regional geographies. Certain collections of papers–festschriften and other edited selections of original or reprinted essays–and conference proceedings are also discussed. In addition to these forms of literature, included here are two important aspects of geography which do not fit conveniently into the divisions used in subsequent chapters, namely applied geography and geographical teaching.

Although not conceived as general introductions to geography, and not at all suitable for the beginning student, there are several books which explore the whole philosophical and conceptual relevance of geography as a discipline by analysis of our perception of and attitudes to the environment. This is a distinctive (but certainly not entirely new) approach to the uniquely geographical concept of environmental synthesis, and can only be considered, therefore, in the context of general geography. The content of such books ranges broadly across the whole field and they introduce, indirectly and at an advanced level, the essential nature and purpose of geography and its basic assumptions.

Perhaps the most systematic treatment of this type is *Topophilia: a study of environmental perception, attitudes, and values* by Yi-Fu Tuan (Prentice Hall, 1974), but this is a difficult book to read. Equally stimulating, and rather more readable, is *The experience of landscape* by J Appleton (Wiley, 1975), which although written by a geographer specifically claims not to be a geography book! In fact it draws on a very wide range of material from various disciplines in a discussion of the 'aesthetics of landscape', relating landscape patterns to human behaviour, attitudes, and experience. A third example, with an expressive title, is

Environmental interaction: psychological approaches to our physical surroundings edited by D Canter and P Stringer (University of Surrey Press, 1975). Although much of this volume is concerned with technological aspects of the architectural environment, there is plenty of material of interest to the geographer and the subject entry for this book in the Library of Congress catalog is 'Man–influence of environment', which comes very close to one of the traditional definitions of geography. *Geographies of the mind: essays in historical geosophy in honour of J K Wright*, edited by D Lowenthal and M J Bowden (Oxford U P, 1976) is a collection of eight essays on the theme of environmental perception which serves to illustrate that this is not in fact a new area of interest; it was an important facet of Wright's work a number of years ago, and his pioneering contributions in this area are listed in a bibliography of his published writings included in this memorial volume.

Introductory texts

This group of books falls into two well-defined categories: firstly there are a number of concise general introductions to geography intended to explain briefly the scope and character of the subject, the major areas of specialisation within it, its relevance and relationships to other disciplines, and chiefly as a guide to prospective students. Perhaps the most satisfactory of these, certainly in the context of the last-mentioned function, is *Geography: an outline for the intending student*, W G V Balchin (Routledge and Kegan Paul, 1970). This book emphasises the uniqueness of geographical methodology and includes information on geography in British universities and career prospects for qualified geographers. A very similar approach is followed in an American volume of comparable scope entitled simply *Geography*, edited by E J Taaffe (Prentice Hall, 1970). At a more elementary level, suitable for students in the later stages of secondary education, is *Background to geography* by D Money (University Tutorial Press, 1975). *Techniques and concepts in geography: a review* by M T Daley (Nelson, 1973) emphasises methodological issues and was written as a corrective to the stress laid in recent years on quantitative approaches, but also serves in a wider role as a general introductory text. *Guide de l'étudiant en géographie* by A Cholley (Presses Universitaires de France, 1942) has been described as 'one of the clearest statements on the scope and purpose of geography and its several branches in any language'.

A pattern established by several works within this group is that of a systematic survey of geographical specialities, typically prefaced by introductory historical or philosophical discussion of the status and nature of geography as a whole. *Background to geography* by G R Crone (Museum Press, 1964, reissued in paperback by Pitman, 1968) is 'an attempt to interest the general reader in the study of geography, and to explain the value of a geographical outlook on the world today'. It concludes with several case-studies. *An introduction to advanced geography*, E W H Briault and J H Hubbard (second edition, Longmans, 1968) is a more factual account, reflecting its origins as a sixth-form text book designed to bridge the gap between school geography and more advanced studies. *The spirit and purpose of geography*, S W Wooldridge and W G East (revised edition, Hutchinson, 1966), is a very readable and well-tried text which has been revised and reprinted many times since its first appearance over twenty years ago. It provides a very convenient and concise review of all the major divisions of the subject, with suitable examples, but not over weighted with superfluous detail. *The scope of geography* by R Murphey (second edition, Rand McNally, 1975) is basically an abridged version of *An introduction to geography* (see below). Starting from a discussion of geography within the social sciences, the book surveys the concepts of region and environment with particular emphasis on the physical environment, and returns finally to the human use of the resources which the environment provides.

Geography: its scope and spirit by J O M Broek (Merrill Books, 1965) is one of a series of books devoted to subjects within the social sciences designed to relate the progress of continuing research to the needs of teaching. It is a most useful distillation of recent thinking in geography into a convenient and readable format. A similar intention is evident in *The changing nature of geography*, R Minshull (Hutchinson, 1970). A major theme is the necessity for precise definition of the unique nature of the subject, while pointing out the ways in which geography is extending its frontiers into territory shared by specialists in other disciplines. A less direct method of definition is followed in *The geographer's craft* by T W Freeman (Manchester University Press, 1967). After an introductory statement summarising the main themes and techniques of geographical study, the approach is historical and biographical. The work and interests of several scholars of different periods and different geographical specialities are discussed in turn as examples of geography in practice.

Handbooks

The second group of general texts are characteristically more factual, much more detailed as statements on the subject content of geography, and almost exclusively American in origin. The function of these volumes goes beyond the definition and introduction of the subject and an outline of its components: rather they are factual summaries of essential information basic to each branch of the subject. They are informative rather than indicative.

Elements of geography by G T Trewartha, A H Robinson, and R H Hammond (fifth edition, McGraw Hill, 1967) is one of the best known of this group. The organisation of the volume is systematic, proceeding from the physical to the cultural and human aspects of geography. An identical pattern is found in *Introduction to geography* by H M Kendall, R M Glendinning, and C H MacFadden (fifth edition, Harcourt Brace, Jovanovich, 1976), *Essentials of geography* by O W Freeman and H F Rays (second edition, McGraw-Hill, 1959), and *Geography: a study of its elements* by Q T Moran and W Moran (Oxford U P, 1969). The volume by R Murphey already referred to above, *An introduction to geography* (second edition, Rand McNally, 1966) is similar but includes extensive regional chapters, as do *The earth and you*, N J G Pounds (Murray, 1962), and *Geography in world society*, A H Meyer and J H Streitelmeier (Lippincott, 1963).

H J de Blij has contributed two handbooks which are worth a mention here. The older example is *Geography: regions and concepts* (Wiley, 1971), which emphasises certain key systematic themes within the context of a general regional framework. More recently, in *Man shapes the earth: a topical geography* (Wiley, 1974), he has used a straightforward topical approach with ten fairly self-contained and well-illustrated chapters on major specialisms. Unfortunately there are no bibliographies. A more philosophical and theoretical work, but otherwise similar in intent, is *Spatial organisation: the geographer's view of the world* by R Abler, J S Adams, and P Gould (Prentice Hall, 1971). Almost unique among these texts in that it is by a British geographer, although it is very obviously aimed at the North American market, is *Geography: a modern synthesis* by P Haggett (Harper and Row, 1972). This 'tries to synthesise at two levels: first, by bringing together the different traditions and themes within the field; second, by stressing the synthesising role of geography as a whole in relation to neighbouring fields'.

It is difficult to differentiate between these handbooks. In both content and arrangement they are essentially very similar, and are designed as background introductory reading suitable for the beginnings of undergraduate courses in geography. An alternative use of such handbooks, in the absence of an English language encyclopedia of geography, is as reference sources of factual information, although none of these are really indexed adequately for this purpose.

Collections

A perhaps less direct, but most valuable approach to the recognition and understanding of the many facets of geography, is via the considerable and increasing number of collections of essays. In some cases specially commissioned and in others reprinted from a wide range of journals and brought together in illustration of a chosen theme or in honour of a particular scholar, such volumes have assumed an important place in the literature of geography. An informal and very readable selection of papers entitled *Let me enjoy* is by O H K Spate (Methuen, 1966) and contains several interesting contributions relevant to the nature of geography. Another volume of reprinted papers, some of which are among the classics of geographical literature, is *Outside readings in geography* edited by F E Dohrs, L M Somers, and D R Petterson (Crowell, 1955). One of the stated aims of the book is to facilitate 'an approach to the study of geography through synthesis of discreet materials rather than through more commonly used channels'. *Geography as human ecology: methodology by example* edited by S R Eyre and G R J Jones (Arnold, 1966), is a collection of case studies illustrative of the same theme, brought into perspective in an introduction by the editors. A much larger collection, similarly suggestive of an ecological framework for the subject, is *Water, earth, and man: a synthesis of hydrology, geomorphology, and socio-economic geography* edited by R J Chorley (Methuen, 1969). Starting from the premise that the study of water provides a logical link between an understanding of physical and social environments, this volume explores the complex relationships and interdependence between physical and human geography. *Man's role in changing the face of the earth* edited by W L Thomas (University of Chicago Press, 1956), consists of the papers presented at a conference organised by the Wenner-Gren Foundation for Anthropological Research. Representing a wide range of interests both within geography and in related fields, this is an important volume again illustrative of an ecological approach. Although published in 1956 the content of the majority of the papers is still remarkably pertinent to modern geography. There are extensive bibliographies accompanying each essay.

Several collections of essays have been planned as stocktakings of the progress of geographical research. The papers in these volumes present an account of the current status and trends of research in the various sub-disciplines of which geography is composed. *Geography in the twentieth century: a study of growth, fields, techniques, aims, and trends* edited by G Taylor (third edition, Methuen, 1957 and later reprints), is an example which has over the years since its first appearance in 1951 become accepted as a standard work. The history and philosophy of geography, and many of the systematic branches of the subject are each discussed by specialists. A more recent volume is *Trends in geography: an introductory survey* edited by R U Cooke and J H Johnson (Pergamon, 1969). Not all fields of the subject are covered in the twenty-six essays which comprise this review, but a wide range of interests is represented, including topics in physical, human, applied and regional geography. Each chapter is written by a specialist, and a particularly useful feature is the extensive references to the literature. A stocktaking volume which is admitted by its editor to have failed in this role is *Directions in geography*, edited by R J Chorley (Methuen, 1973). This was designed to put the main ideas of the quantitative revolution into perspective, but in fact the fourteen contributors, who include many of the leading figures of the last few years, offer more insights into the likely direction of future research than into the recent past.

Some collections assembled as memorial volumes reflect the interests of certain individual geographers to whom they are dedicated; others are deliberately and carefully balanced to include contributions from contrasting aspects of the subject. There are a considerable number of these collections published, and some, for example *Liverpool essays in geography* edited by R W Steel and R Lawton (Longmans, 1967), have achieved the status of standard works. Others are notable as illustrations of the contributions of particularly influential schools of geography, such as *Processes in physical and human geography: Bristol essays* edited by R Peel, M Chisholm, and P Haggett (Heinemann, 1975). This contains a well balanced collection of twenty research papers by Bristol geographers to mark the fiftieth anniversary of the establishment of the University of Bristol Department of Geography. However, the scope of most examples differs so much that generalisations and reference to particular volumes are not helpful. In most cases the detailed contents are listed in the abstracting and indexing journals, and the individual essays are more important than the collections as such.

This is true also of papers presented at conferences and symposia organised by national and international societies and other groups of geographers, and subsequently published. The International Geographical Congress, held usually every four years, is perhaps the most obvious example, although recently the trend has been to publish only volumes of abstracts, in contrast to the former multi-volume sets of complete proceedings. The importance of such papers, as has been stated elsewhere, is that they represent a relatively current statement of existing knowledge, with an immediacy more comparable to journal publication than to books. As in the case of other forms of collected works, however, it is the individual contributions which are of significance rather than the collections themselves. For this reason they are most frequently approached via indexing and abstracting journals, in which the constituent papers from individual symposia are normally listed. A second useful source is by way of other publications of the bodies under whose auspices such meetings are held. This is particularly true of national geographical socieites, most of which organise regular conferences: thus, for example, abstracts of conference papers form a regular feature in the *Annals of the Association of American Geographers*, and a similar function is performed in the UK by the *Transactions of the Institute of British Geographers*, where papers presented at annual meetings are printed in full.

Applied geography

Geography as a field of academic study was a relatively late arrival. Since becoming firmly established in the universities, however, there has been a continuing tendency towards self-justification, not only in terms of scientific standards and stature as a viable body of theory and methodology, but in relevance to social, human, and technical problems. The term 'applied geography' summarises this movement embracing every aspect of the subject capable of practical application. Although both the phrase and the concept are found much earlier, applied geography was at its most fashionable in the 1960s. The concept is still very much present in modern geography, but the terminology has changed. Much current literature on environmental problems such as conservation, pollution, population, and physical and resource planning falls within the field, but has tended to make it a less coherent whole. A number of books typifying these trends will be discussed later as human geography; there are however a few general introductions to the idea of geography as an applied science which it is appropriate to mention here.

Professor L D Stamp's *Applied geography* (Penguin, 1960) was largely responsible for the popularisation of the concept in this country. As organiser of the first land utilisation survey of Great Britain, and later as director of the IGU's world land use survey, Stamp had wide experience of the practical application of his subject, and a considerable portion of his book is devoted to land use problems; further chapters cover town and country planning, industry, trade, and so on. Stamp's contribution was recognised in a memorial volume of essays published in 1968 by the Institute of British Geographers, developing similar themes, entitled *Land use and resources: studies in applied geography.*

A more philosophical but very readable introduction is *From geography to geotechnics* by B MacKaye, and edited by P T Bryant (University of Illinois Press, 1968). This consists of essays previously published in a variety of journals and 'selected as representative works of continuing interest to planners, scholars, conservationists, and private citizens concerned about the effects of our urbanised and deteriorating environment on the quality of our lives'. A rather different approach is found in a pamphlet produced by the Association of American Geographers and the National Council for Geographic Education, *Geography as a professional field* edited by P E James (second edition, 1971). Although teaching has long been the largest source of employment of professionally trained geographers, increasing numbers are entering the various fields of applied geography. The book is a review of career prospects in these fields, ranging from military geography to agriculture.

A more orthodox introduction to applied geography is a volume published in connection with a meeting of the Canadian Association of Geographers, entitled *The geographer and society* edited by W R D Sewell and H D Foster (University of Victoria, British Columbia, 1970). A general discussion of the relevance of geographical research is followed by a series of essays on the application of different geographical concepts in various fields, including a number of case-studies. A comparable British volume is *Evaluating the human environment: essays in applied geography* edited by J A Dawson and J C Doornkamp (Arnold, 1973), which contains eleven widely ranging studies covering most of the main themes and concerns of applied geography: the addition of a substantial introduction to the philosophical context of applied geography within the discipline would make this a most useful general text. *Geography and contemporary issues: studies of relevant problems* edited by M Albaum (Wiley, 1973), consists mainly of reprints from non-geographical journals, organised under the main headings of Poverty, Race relations, Urban blight, Population, Environment, and Conflict.

The theme of relevance is made the principal justification for all geographical study and applied geography is used as a selling point to prospective students in *An invitation to geography* edited by D A Lanegran and R Palm (McGraw-Hill, 1973). More restricted in scope are *Geography and man's environment* by A and A H Strahler (Wiley, 1977), which relates the study of physical geography and ecology to environmental issues and problems; and *Natural hazards: local, national, global*, edited by G F White (Oxford U P, 1974), which contains thirty-two specific case studies by over fifty authors.

There are several useful reviews of applied goegraphy in the French language of which three are discussed below. *Géographie et action: introduction à la géographie appliquée*. M Philipponeau (Colin, 1960), explores the origins of applied geography and its practice in various countries, emphasising its role in French geography. The second part consists of a brief review of the main branches of the concept. A similar approach is followed in *La géographie active* by P George, R Guglielmo, B Kayser, and Y Lacoste (Presses Universitaires de France, 1964). The third work is *Les géographes au service de la société* by J A Sporck and O Tulippe (Comité National de Géographie, Royaume de Belgique, 1968). In two parts, this consists of a study of applied geography as practised in Belgium and abroad.

Finally, reference should be made to a volume of conference proceedings, which present the most comprehensive summary of the application of geographical techniques and concepts throughout the world. This conference, Colloque International de Géographie Appliquée, Liège, 1967, was organised by the Commission on Applied Geography of the International Geographical Union, and the proceedings edited by C Christians. Papers are in English and French, but represent a very broad spectrum of national points of view.

As previously stated, much of the literature is devoted to specific aspects of applied geography, and these more detailed studies are described elsewhere.

Geography teaching

The teaching of geography at all levels is a topic which has given rise to a considerable volume of literature. A high proportion of trained geographers eventually take up careers in the academic teaching of the subject, and geography is widely taught both as an element within liberal studies programmes and as a field for specialisation. Broadly speaking, books on this subject fall into two categories: firstly, there are a number of texts which discuss such matters as the philosophical basis of geography as an educational discipline, its place in the curriculum, and methods and techniques suitable for geographical instruction; and secondly, there are several sourcebooks designed to guide the teacher to sources of documentation, statistical information, and details of teaching aids and suppliers of these.

The teaching of geography by G H Gopshill (third edition, Macmillan, 1966) is primarily concerned with the methodology of geography teaching; the use of illustrations, maps and practical assignments, and the problem of relevance. The book concentrates on general principles and avoids detailed discussion of course content. *Methods of geographic instruction* edited by J W Morris for the National Council for Geographic Education (Blaisdell Publishing Company, 1968), is more comprehensive. Twenty-eight essays analyse the philosophy and purpose of geography, the use of teaching aids, geographical methodology both in regional and systematic studies, and the contribution of geography to education. *Frontiers in geographical teaching* edited by R J Chorley and P Haggett (second edition, Methuen, 1970) is a collection of papers originating in a conference for teachers of geography. Several of the papers discuss the teaching of geography at various levels, but the greater portion of the book is taken up with essays on new concepts and techniques in different branches of the subject. Each section concludes with a short bibliography, and in its capacity as a summary of recent developments is of considerable interest in a wider context than that originally conceived.

There have been several recent thoughtful contributions to the theoretical and philosophical background to geography teaching. *Geography in education* by N J Graves (Heinemann, 1975), and *Geography and the geography teacher*, D Hall (Allen and Unwin, 1976) are typical, and are written with very similar objectives in mind. They both contain more fundamental discussion than that found in more general, practical, handbooks, and relate geography to other subjects in the school curriculum. *Evaluating the geography curriculum* by W E Marsden (Oliver and Boyd, 1976) aims to relate curriculum theory to geography teaching, as does *Geography and the integrated curriculum: a reader* (Heinemann, 1976), edited by M Williams and containing twenty-eight contributions (all British) dating from as early as 1909.

There have been several attempts to challenge the traditional scope of geography as a school subject and to suggest a broader approach. Notable among these are *Insights into environmental education*, edited by G C Martin and K Wheeler (Oliver and Boyd, 1975), which is a varied and stimulating collection of papers; and *Geography for teachers: perspectives in geographical education* edited by J Bale, R Walford, and N J Graves (Oliver and Boyd, 1973), which contains twenty-six articles, many of which are deliberately provocative, reprinted from a range of journals.

More orthodox examples are *Geography in and out of school*, E W H Briault and D W Shave (second edition, Harrap, 1967) and *The teaching of geography in secondary schools* (fifth edition, Cambridge U P 1967). The latter work, issued by the Incorporated Association of Assistant Masters in Secondary Schools, is a well established standard text which originated in a *Memorandum on the teaching of geography* originally published in 1935. Philosophical considerations are incorporated as well as practical guidance on equipment, examinations, methods of work, and teaching technique. *Place and people: a guide to modern geography teaching* by S Dunlop (Heinemann, 1976) is a very concise but useful guide to current practice. Teaching in the lower age ranges is covered in *Geography and younger children: an outline of theory and practice* by E J Barker (University of London Press, 1974) and in the more traditional *Fundamentals in school geography: a book for teachers and students in training* by O Garnett (third edition, Harrap, 1965, and reprinted). *Teaching geography* by D Bailey (David and Charles, 1974) is a very readable and sensible introduction to secondary teaching, while more advanced work is discussed in *The teaching of geography at university level*, J Tricart (Harrap, 1969), one of a series of books entitled 'Education in Europe' sponsored by the Council for Cultural Cooperation of the Council of Europe. This is a report on the current state of affairs, together with conclusions on the effectiveness of geographical education and recommendations for the future. *The social sciences and geographic education: a reader* edited by J M Ball, J E Steinbrink, and J P Stoltman (Wiley, 1971) is a collection of papers on philosophical, psychological, and methodological aspects of geography teaching, considering geography in relation to its broader implications in other social science fields. Another useful collection is *New movements in the study and teaching of geography*, edited by N Graves (Temple Smith, 1972). The essays in this volume are about equally divided between statements on the changing nature of geographical concepts and their use in the classroom, and discussion of various problems of teaching and learning. The emphasis on the psychological aspects of this process distinguishes this volume from most of those mentioned above.

The specialisation of certain journals on geographical teaching (notably *Geography* and *Journal of geography*) has been noted in the previous chapter. But the Geographical Association and the National Council for Geographic Education, which are respectively the publishers of these two journals, both publish also a range of pamphlets and short monographs on specialised aspects of the subject, including bibliographies of textbooks, eg *Geography books for sixth forms* by L J Jay (Geographical Association, 1971–Teaching geography, number 16). Another useful example is *Geography in education: a bibliography of British sources 1870-1970*, by C T Lukehurst and N J Graves (Geographical Association, 1972); this is a select and unannotated guide to both books and journal articles, arranged by subject. The National Council for Geographic Education also sponsors more comprehensive monographs such as *Methods of geographic instruction* (discussed above) and *A handbook for geography teachers*, edited by R E Gabler (1966). The best way of keeping abreast of the output of both of these organisations is by use of their journals.

Another work bearing the same title, *Handbook for geography teachers* (sixth edition, Methuen, 1971) was prepared by the Standing Subcommittee in Geography of the London University Institute of Education. Originally published in 1932, the current edition is edited by M Long. A large proportion of the book is taken up with lists of texts on various aspects of geography, suitable for use in schools in conjunction with orthodox syllabi. In addition there are lists and descriptions of field work centres, atlases and maps (including wall maps), visual aids, publishers, and embassies and commercial sources of information and teaching materials. A further source of similar information is *The geographer's vademecum of sources and materials*, J C Hancock and P F Whiteley (new edition, Philip, 1977). Arranged in two sections by topics and regions, this concise handbook does not include lists of books. Documentary sources do however figure in *Unesco source book for geography teaching*, published in 1965 by Longmans in association with Unesco. In common with several of the books on geography teaching already described, this includes chapters on the importance and nature of geographical education, teaching techniques and materials, the syllabus, and the geography room. The bibliography includes a number of entries relating to the teaching of geography in various countries of the world.

Six

CARTOBIBLIOGRAPHY

THE TRADITIONAL and continuing importance of maps in geographical study makes it imperative that the geographer be familiar not only with the main sources but is able to identify and locate more specialised items when necessary. Tracing published maps and atlases retrospectively and keeping abreast of current publication in this area are the main themes of this chapter. Cartography as a field of study is discussed elsewhere (chapter ten), and cartobibliography is defined here as the listing of cartographic production, concerned with published maps and atlases and the bibliographical apparatus for controlling these. To some extent this distinction is artifical, especially in terms of historical cartography, in that the main histories are not only about maps but are also sources on what is available; the same is true, to a lesser extent, of some texts dealing with modern cartographic techniques.

Another distinction implied here, which is also slightly spurious, is that between sources of information on books and on maps and atlases. In fact, some of the general bibliographies of books–including many national bibliographies–do contain a rather wider selection of materials. Naturally enough, coverage in such sources is more effective for atlases than it is for sheet maps. However, the most comprehensive and most reliable cartobibliographic sources are those which specialise, and it is the specialised sources which are our concern here. It is of course quite logical that the distinction should exist; generally speaking, maps and atlases are produced by a relatively small number of publishers (many of whom publish little else), the main library collections are specialised ones, and even in general libraries they are usually kept separate for obvious physical reasons. Beyond the various general bibliographies there is a wide range of sources concerned solely with maps and atlases, and these form a second, parallel, bibliographic structure; within this separate structure, there are examples of most of the several categories which have already been described for books. Cartobibliography is thus both an integral part of the literature of geography and a unique, specialised, and complex field in its own right.

General guides

There is no totally satisfactory general guide to this 'literature'. The nearest approach to such a guide is *Modern maps and atlases: an outline guide to twentieth century production* by C B M Lock (Bingley; Hamden (Conn), Linnet, 1969), but this does not deal with historical cartography. It describes both the techniques of production and discusses in some detail a wide range of national, regional, and thematic maps and atlases. It is in narrative form, well supported by a very full index. As a source of information on what is now available, its usefulness is limited by its age, and Dr Lock has chosen to incorporate a revision in her *Geography and cartography: a reference handbook* (same publishers, 1976) rather than producing a second edition limited to cartography. In its new form it serves a quite different function as a quick reference source, and the information on maps and atlases is dispersed under alphabetically arranged headings covering a far wider field. A more coherent account of the sources is available in the first few chapters of *Map librarianship*, by H Nichols (Bingley; Hamden (Conn), Linnet, 1976) although this is a difficult book to read. A good selection of the main catalogues and indexes are described, with helpful critical comment, and the book covers both historical and modern maps and atlases. Many useful specialised bibliographies, as well as bibliographical articles and shorter notes, are available in the *Bulletin* of the Geography and Map Division, Special Libraries Association, published quarterly since 1950. A cumulative index has been produced to facilitate tracing these for the period 1950-1967.

As with books, the principal library collections constitute a major source of information on what is available, and some of the published catalogues and accessions lists of these are discussed below. The main collections themselves are detailed in a recently published *World directory of map collections* compiled by W W Ristow and sponsored by the Geography and Map Section of the International Federation of Library Associations (Verlag Dokumentation, 1976). Failing this, there is a listing in *Orbis geographicus*, which unfortunately amounts to only 129 of the largest collections. Again, only brief details are given, sufficient to identify each institutional head and postal address, but no indication of the scope of the collections or of special strengths. For North America there is a much more comprehensive directory prepared by the Geography and Map Division of the Special Libraries Association. *Map collections in the United States and Canada: a directory* (second edition, 1970) lists over six hundred collections, arranged by geographical location. Map collections are included to some extent in the general directories of libraries described in chapter two, but the coverage here is far from comprehensive.

The most useful source for many purposes, and indispensible as a guide to materials currently available for purchase, is *International maps and atlases in print* edited by K L Winch (second edition, Bowker, 1976). A sample page from this is reproduced as figure thirteen. This includes both official productions and maps from commercial sources and is genuinely international in its scope, although some specialised map forms (such as marine navigation charts) are excluded. It is arranged under countries, and within this by a simple scheme of broad subject divisions, with maps and atlases being listed separately. There are over eight thousand entries and about seven hundred different publishers are represented. Hopefully new editions will appear at fairly frequent intervals.

There is a major bibliography of atlases produced by the Library of Congress Map Division, which lists over 18,000 items, most of which are now inevitably of mainly historical interest. The original work was published in four volumes between 1909 and 1920 under the title *A list of geographical atlases in the Library of Congress, with bibliographical notes*, compiled by P L Phillips and is available also in a reprint edition by Theatrum Orbis Terrarum, 1970. Three supplementary volumes listing more recent acquisitions and compiled by C E LeGear were published in 1958, 1961 and 1973, and an index volume in 1974. This unique and valuable work of reference may be supplemented by *Guide to atlases: world, regional, national, thematic; an international listing of atlases published since 1950*, G L Alexander (Scarecrow Press, 1971). There are no annotations but detailed indexes (publishers, languages, authors, cartographers, and editors) are included. *Official map publications: a historical sketch and a bibliographical handbook of current maps and mapping services in the United States, Canada, Latin America, France, Great Britain, Germany, and certain other countries* compiled by W Thiele (American Library Association, 1938) was a major guide to map production at the date of its publication. It provides a useful historical perspective but is of little use for current work; a revised and updated edition would be an extremely valuable addition to the literature. The official map publications of the United Kingdom are covered in *Ordnance Survey maps: a descriptive manual* by J B Harley, published by the Ordnance Survey itself in 1975. This includes very full explanations and

597 Vietnam

See also: 596/598 INDO-CHINA
59 SOUTH EAST ASIA

A1 GENERAL: ROADS

BAN DO VIETNAM - VA DUONGSA
1:2,000,000
Dalat : NGD, 1969
70 x 100
Road map with physical colouring, province boundaries. Legend in Vietnamese.

EAST ASIA ROAD MAPS
Series 1306
1:1,250,000
AMS Series 1306
Washington : AMS, 1968
81 x 74 each
2 sheet road map, contours, layer tints, roads classified acc. surface, railways, vegetation, boundaries:
1 North Vietnam and adjacent territories
2 South Vietnam

REPUBLIC OF VIETNAM ROAD MAPS
1:250,000
In 9 sheets
Dalat : NGD, 1966
Road maps series covering South Vietnam only, available flat or folded in case. Roads classified, spot heights, railways, vegetation, distances in km. Restricted release.

B TOWN PLANS

CITY MAPS
Series L909
1:15,000
Dalat : NGD
Restricted release maps available for
Saigon, 1968
Kontum, 1961

CITY MAPS
Series L909
1:12,500
Dalat : NGD, 1963+
Series of 1 sheet maps, on restircted release.
Available for:
Khanh Hung
Rach Gia
Long Xuyen
Chau Doc
Yung Tau
Phan Thiet
Phan Rhang
Dalat
Nha Trang
Ban Me Thuot
Qui Nhon
Pleiku
An Khe (An Tuc)
Quang Ngai
Chu Lai
Da Nang
Hue
Quang Tri

CITY PICTO MAPS
Series L909
1:12,500
Dalat : NGD, 1969+
Restricted release maps available for:
Can Tho
Cu Chi
Phu Bai
Quang Tri
Gio Linh
Dong Ha

CITY MAPS
1:10,000
Dalat : NGD
Restricted release maps available for:
Can Tho, 1967, 2 sheets
Vinh Long, 1969
My Tho, 1966
Do Thanah Saigon, 1959, 2 sheets
Saigon - Tan Son Nhut - Gia Dinh, 1966, 6 sheets
Tay Ninh, 1967, 2 sheets
Bien Hoa, 1968, 4 sheets
Dalat, 1960
Dalat, 1967, 4 sheets

TOPOGRAPHIC CITY MAPS
1:5,000
Dalat : NGD
Plans available: Dalat City, 1952
Ban Me Thuot City, 1951

C OFFICIAL SURVEYS

INTERNATIONAL MAP OF THE WORLD
1:1,000,000
AMS Series 1301
7 sheets needed for complete coverage.
see WORLD 100C and index 100/1.

VIETNAM ADMINISTRATIVE AND ROAD MAP
1:500,000
In 12 sheets, see index 597/1
Dalat : NGD, 1965-6
100 x 60
Col. maps, relief shading, communications, vegetation, land use.

TOPOGRAPHIC SERIES L701
1:50,000
In 418 sheets
Dalat : NGD, 1965+
Restricted release series, contoured in cols.

TOPOGRAPHIC MAP SERIES L8015
1:25,000
In 51 sheets
Dalat : NGD, 1967+
Restricted release series.

PICTOMAPS - Series L8020
1:25,000
In 527 sheets,
Dalat, NGD, 1966
Restricted release maps covering South Vietnam.

ORTHOPICTOMAPS
Series L8023
1:25,000
Dalat : NGD, 1970+
Restricted release series.

TOPOGRAPHIC MAPS
Series L8021
1:10,000
Dalat : NGD
Available on restricted release for:
My Tho - Tan An, 23 sheets
Tau, 9 sheets, 1970
Tay Ninh, 9 sheets, 1969

TOPOGRAPHIC MAPS
1:5,000
In 24 sheets
Dalat : NGD
Series available for various areas, still being printed. Restricted release.

D POLITICAL & ADMINISTRATIVE

V'ETNAM
1:2,000,000
Moskva : GUGK, 1973
Col. study map, showing water features, settlements, admin. centres and divisions, communications, boundaries, economic details, pop, Geog. description.
Ro 0.30

REPUBLIC OF VIETNAM ADMINISTRATIVE MAP
1:1,000,000
Dalat : NGD, 1967
78 x 105
Shows admin. divisions of South Vietnam and pop. figures. List of admin. areas on reverse.

ADMINISTRATIVE AND ROAD PROVINCE MAPS
Various scales
In 44 sheets
Dalat : NGD, 1971
Showing admin. boundaries and roads classified. 1 sheet for each province at scales of 1:100,000, 1:150,000, 1:200,000, 1:250,000

E1 PHYSICAL: RELIEF

WIETNAM
1:4,000,000
Warszawa : PPWK, 1968
43 x 46
Relief shading, communications, vegetation
Text in Polish.

V'ETNAM - SPRAVOCHNAYA KARTA
1:2,000,000
Moskva : GUGK, 1968
58 x 90
Reference map with physical col. contours, communications, boundaries. 5 thematic insets, 20pp text and index in Russian.

VIET NAM
1:1,000,000
Washington : US Air Force
91 x 172
Physical - political map, contours, 3-D relief shading; for identifying war areas.

Figure 13: A sample page from 'International maps and atlases in print'

Japan Tachikawa

Tachikawa-shi gaizu [Street map of Tachikawa]. 1:10 000. Toshi-chizu shirizu [City Map Ser.] C31302. Ōsaka [etc.]: Shōbunsha, 1971. Col. map 74.5 × 51 cm, fold. 10.5 × 21.5 cm

On verso: Toshi keikaku gairomō zu [Map of urban planning road network] (overprinting in red on outline base map of recto)
In Japanese

Japan Takamatsu

Tachikawa-shi gaizu [Street map of Tachikawa]. 1:10 000. Toshi-chizu shirīzu chizu shirīzu [City Map Ser.]. Osaka: Shōbunsha, 1971. Col. map 82.5 × 59 cm, fold. 10.5 × 22 cm

On verso: Takamatsu-shi zenzu [General map of Takamatsu City district]; 1:50 000; 54 × 58.5 cm
Ya-shima; 1:20 000; 19 × 32 cm
Ritsurin kōen [Ritsurin Park]; 1:3 300; 19 × 26 cm
With accompanying text 24 p.: ill.; 22 × 11 cm

Lebanon

Carte du Liban. 1:20 000. Établie par l'Armée Libanaise. Beyrouth: Direction des Affaires Géographiques, Ministère de la Défense Nationale, 1971–72. 5 col. maps 57.5 × 46 cm

H–8 Ouadi El Qarn
N–11 Atneine
Q–12 El Qâa (Station)
R–12 Joussié
S–11 En Nsoûb

Mongolia

Mongol'skaya Narodnaya Respublika. 1:3 000 000. [Karta sostavlena . . . Nauchno-redaktsionnoy Kartosostavitel'skoy Chast'yu GUGK v 1966 g. Ispravlena v 1972 g.]. [Vtor. izd.]. [Redaktory N.I. Arel'yeva i L.M. Voronina.]. [Moskva: Glavnoye Upravleniye Geodezii i Kartografii, 1972]. Col. map 81 × 47 cm, fold. 14 × 23 cm

Insets: Ekonomicheskaya karta, Klimaticheskaya karta; 1:12 000 000; 20 × 11 cm
With accompanying notes & gazetteer by D. A. Chumichev.– 27 p.; 20 cm

Philippines

[The Philippines]. 1:250 000. S–2500. Jayme V. Presbitero Director of the P.C.G.S. Manila: Philippine Coast and Geodetic Survey, 1976. 4 col. maps 64 × 44 cm or smaller

2508 Solano. Ed. 3 May 1976
2509 Tarlac. Ed. 3 Jan. 1976
2526 Taytay. Ed. 3 May 1976
2536 Brokkes Point. Ed. 3 July 1976

AFRICA

Africa South Gazetteer

Southern African place-names/O. A. Leistner & J. W. Morris.– Grahamstown: Albany Museum for the Cape Provincial Museums, 1976.–[v], 565 p.; 24 cm.– (Annals of the Cape Provincial Museums; vol. 12)

Contains *ca* 42 000 names of places in South Africa, South West Africa, Botswana, Lesotho, & Swaziland
"This list has not been approved by the Place Names Committee or any other official body . . ."– Introduction
The authors are on the staff of the Botanical Research Institute, Dept of Agricultural Technical Services, Pretoria, S.A.

Figure 14: A sample page from 'New geographical literature and maps'

descriptions of all the standard series, with illustrations and lists of supplementary readings.

Current bibliographies

One of the most useful of the guides to current material apart from *International maps and atlases in print* has no direct parallel among the bibliographies of books. This is the annual *Geo-Katalog.* It includes maps, atlases, globes, and guidebooks from publishers all over the world, arranged in a regional sequence and supported by a thematic appendix containing geological maps, aeronautical charts and other more specialised items. It is by nature a sale catalogue, and is restricted to items 'in print' at the time of publication. A more orthodox current bibliography—indeed the only official international listing—published annually since 1949, is the *Bibliographie cartographique internationale.* Until 1970 this was published by Armand Colin for the Comité National Français de Géographie and sponsored by the International Geographical Union; more recently the compilation has been undertaken at the Bibliothèque Nationale and the publication has been from the Librairie de la Faculté des Sciences. The first volume listed maps published in 1947-8 in eight collaborating countries; more than twenty leading geographical societies from around the world now contribute, and recent editions have contained well over three thousand entries. Publication however is rather irregular, and about two years in arrears. Arrangement is regional, but the whole is elaborately and fully indexed.

Where these specialised sources are not available, the best alternative for information on current cartographic production is to use the lists in general geographical indexing journals. The best of these for this purpose is *New geographical literature and maps*, published twice yearly by the Royal Geographical Society. This lists additions to the society's map room in a regional arrangement, with systematic subdivision where appropriate: no indexes are provided. The corresponding American publication, *Current geographical publications*, has included a separate section on maps since 1964; before this time they were incorporated among entries for books and periodical articles. Arrangement is mainly regional, and the lists represent a selection of the more interesting and important additions to the collections of the American Geographical Society. There are ten issues per year. Other major map libraries also produce lists of their accessions, and these are of considerable importance, particularly in the case of collections based on legal deposit; good examples are the Bodleian Library Map Section's *Selected book and map accessions*, and the *Catalogue of printed maps accessions* of the British Library Map Library. The general national bibliographies of many countries include maps and atlases, though not always as separate lists, and the coverage is not usually very good, particularly in the case of sheet maps; Nichols' contains a list of these.

Geographical journals play an important role in reviewing major maps and atlases as they are published, and provide a useful supplement to the sources detailed above. Several of the major titles have sections specifically devoted to cartographic news and announcements: the *Geographical journal*, for instance, includes a 'Cartographic survey' which contains reviews of new maps and atlases as well as books on cartography. Secondly, there are a number of important cartographic journals which include bibliographical information in addition to their other function of providing a forum for technical articles on cartography. The most important of these from a bibliographical point of view are *World cartography* (United Nations Department of Economic and Social Affairs, irregular, 1951-); *Surveying and mapping* (American Congress on Surveying and Mapping, quarterly, 1941-); *American cartographer* (American Congress on Surveying and Mapping, semi-annual, 1974-); and *Cartographic journal* (British Cartographic Society, semi-annual, 1964-).

Finally, there are the lists and catalogues issued by the publishers of maps and by mapping agencies themselves. The greatest number of maps produced come from 'official' agencies in each country, government sponsorship being necessary in view of the high costs and sophisticated organisation required to conduct an accurate national survey. These agencies invariably publish their own catalogues and while most of the information (certainly in the case of the industrialised countries of the western world) is available in such sources as *Geo-Katalog* and *International maps and atlases in print*, these will sometimes have to be consulted direct. The various national agencies are listed in *International maps and atlases in print*, and (with addresses) in a very useful article on 'Published sources of information about maps and atlases' by R W Stephenson in *Special libraries* volume 61, 1970, 87-98, 110-112. This list includes not only the official bodies but also commercial publishers and sellers of maps and atlases. Many of these are well known specialist companies–Bartholomew, Philip, Stanford, Westermann, Ravenstein, Rand McNally–but Stephenson also includes many lesser known publishers and agents in a list covering all major countries.

Retrospective bibliographies

There is an immense variety of retrospective bibliographies of maps and atlases, but they fall into two quite distinct categories. The most comprehensive and most generally useful sources are those based on major collections, rather like the *List of geographical atlases in the Library of Congress* referred to above. Secondly, there are smaller, specialised bibliographies confined to specific periods, places, and themes.

There are two important and very substantial collections to which printed catalogues are available. The holdings of the British Library Map Library (formerly the British Museum Map Room) up to the end of 1964 are listed in the *Catalogue of printed maps, charts and plans*, published by the museum in 1967 in fifteen volumes. This is arranged alphabetically under place names and locations, with additional entries for cartographers, surveyors and editors, and includes books on cartography and globes in addition to the maps, charts, and plans referred to in the title. Secondly, there is the *Dictionary catalog* of the New York Public Library Map Division, published by G K Hall in 1970. This consists of ten volumes and contains entries for nearly 300,000 sheet maps, as well as a large selection of atlases and books on cartography. *Index to maps in books and periodicals*, compiled by the American Geographical Society Map Department (G K Hall, 1968) is not the catalogue of a collection as such but contains details of maps published in a wide range of other literature and which would be very difficult to trace without reference to this unique reference tool. The original publication comprises ten volumes, and a supplement for the years 1968-1971 has also been published. On a very much smaller scale, but nevertheless a useful source of information, is the British National Maritime Museum's *Catalogue of the library:* volume three covers *Atlases and cartography* and was published in two parts in 1971. This contains detailed information on nearly eight hundred atlases, with lists of individual maps.

The great national collections of maps naturally include many examples of unpublished work, and these too are listed in catalogues. The *Catalogue of the manuscript maps, charts, and plans and of the topographical drawings held in the British Museum* was originally published in 1844, and reprinted in 1962. A further example is the Public Record Office *Maps and plans in the Public Record Office*, of which two volumes have so far appeared: 'British Isles, c1410-1860' (volume one, 1967); and 'America and the West Indies' (volume two, 1975). The American equivalent of this work is the *Guide to cartographic records in the national archives* (Government Printing Office, 1971).

J B Harley has produced two useful retrospective guides to British maps. The more recent of these is *Maps for the local historian: a guide to British sources* (National Council of Social Service, 1972). This is a detailed and authoritative description of all the main categories of maps likely to be found in local history collections. In collaboration with C W Phillips he has also written *The historian's guide to Ordnance Survey maps* (National Council of Social Service, 1964), which provides information on all the various series and scales of early maps, including index sheets. This was originally published as a series of articles in the journal *Local historian.* Early English county maps are listed in *The printed maps in the atlases of Great Britain and Ireland: a bibliography, 1579-1870* by T Chubb, originally published in 1927 and reprinted by Dawsons in 1966. (Available in the USA from Shoe String Press.) This lists atlases chronologically and details the individual maps. *County atlases of the British Isles, 1579-1850* by R A Skelton is an attempt to update Chubb's work, but only the first volume, covering the years 1579-1703, has been published (Carta Press, 1970). There are numerous examples of lesser works, listing maps of single counties or towns, characteristically published by local societies, and these may most easily be traced through Harley's 1972 guide cited above. American local atlases are listed in *United States atlases: a catalogue of national, state, county, city, and regional atlases in the Library of Congress and cooperating libraries*, compiled by C E LeGear (Government Printing Office, two volumes, 1950-53). Early American maps are detailed in *Maps and charts published in America before 1800: a bibliography* by J C Wheat and C F Brun (Yale University Press, 1969).

International maps and atlases

There have been several attempts to publish complete sets of sheet maps covering the whole world, perhaps the best known of which is the *International map of the world on the millionth scale*, first proposed in the 1890s and still not complete. Much of the planning was done by a specially appointed commission of the International Geographical Union and the work has been coordinated by the Cartographic Office of the United Nations since 1953. Actual production of the various sheets has been undertaken locally by various national cartographic agencies, working to an internationally agreed specification. Progress in the production of the map is documented in annual reports from the United Nations and new sheets are announced in the *Cartographic journal* and similar publications. The *World map (Karta mira) 1:2,5000,000* was produced between 1964 and 1974 as a cooperative venture between the cartographic agencies of a number of Eastern European countries, each taking responsibility for the preparation of certain areas of the world, including the oceans. While the keys appear in both English and Russian, only Latin lettering is used on the maps themselves and the official forms of names are used throughout. The map consists of 234 sheets, produced to a uniformly high standard. On a scale of 1:500,000 there is the *International topographic map of the world*, produced by the United Kingdom Directorate of Military Survey. Major series such as these are a good standby where more detailed locally-produced maps are not available, providing a most useful supplement to the various national series.

The best general reference atlas available is the *Times atlas of the world: comprehensive edition* (revised edition, the Times, in collaboration with J Bartholomew, 1977). This is a production of excellent quality, well designed and printed, and is a revision of a longstanding work of proven worth. It contains some 240 pages of maps and a full index giving latitude and longitude. The closest rival to the *Times atlas* is the *International atlas* (revised edition, Philip, 1977), produced in association with several foreign publishers. The introductory text and other marginal information is in four languages–English, German, Spanish, and French–and the work is of high quality, though less attractively produced than the *Times atlas*. The number of maps and the scales chosen for each region reflect 'the relative economic and cultural significance on the world scene as well as its total population and area'.

Less comprehensive and less sophisticated atlases include the *Reader's Digest great world atlas* (revised edition, Reader's Digest, 1975), and the *Atlas of the earth* (Mitchell Beazley, 1974). Both of these are encyclopedic in nature, incorporating extensive textual notes and statistical material in addition to the maps. In the latter title this comprises about half the volume. Among a number of excellent atlases from the Oxford University Press the leading examples are the *New Oxford atlas* (1975) and the smaller *Oxford world atlas* (1973) both of which are particularly strong on thematic maps. The maps in the *Atlas of the earth* referred to above are based on those in the *University atlas*, published by Philip and now in its seventeenth edition (1975); this is a good concise reference with clear physical maps and some smaller scale thematic ones. Comparable in scope and also aimed at a student market is another well-established and popular atlas, the Bartholomew *World atlas* (revised edition, 1977). *Goode's world atlas* was originally published in 1922, and is now in its fourteenth edition (Rand McNally, 1974); it includes a pronouncing index of some 35,000 place names. Based on the *Times atlas* is an excellent abridged edition, the *Times concise atlas of the world* (1975) containing 144 pages of physical and political maps.

The above titles are merely a selection from those available and suitable for advanced study. Forty major English language atlases (and a considerable number of lesser examples) are examined in detail in *General world atlases in print: a comparative analysis* by S P Walsh (fourth edition, Bowker, 1973). This contains extensive factual information and critical comment, an assessment of the strengths and weaknesses of each item and an indication of general quality.

National, regional and thematic atlases

General world atlases similar to those cited above constitute a minority of the total range of atlases available. An atlas is simply a collection of individual maps, and the unifying feature may be either a particular locality or a specific subject theme. There are many examples of both types and the discussion below centres on a few typical illustrative samples; the existence of an atlas on any given topic or locality can be checked using the guides and bibliographies already cited.

The compilation of national atlases has been a priority for many countries for a number of years, and was given additional impetus by the establishment in 1956 of a Commission on National Atlases by the International Geographical Union. Such atlases not only provide a valuable source of data for scholarly study and analysis but a sound basis for planning purposes. This is well illustrated by the case of Britain where maps in the national planning series published by the Ordnance Survey on a scale of 1:625,000 constitute part of the official national atlas. More readily recognisable as an atlas, however, is the *Atlas of Britain and Northern Ireland* (Oxford U P, 1963) which is an outstanding example and beautifully produced. Some two hundred pages of maps cover physical characteristics, industry, commerce, transport, agriculture, population distribution and movement, and so on. The atlas provides a complete picture of the United Kingdom and 'is essentially a record of a country's resources at a given point in time'. The *National atlas of the United States of America* (United States Geological Survey) is 'designed to be of practical use to decision makers in government and business, planners, research scholars, and others needing to visualise country-wide distribution patterns and relationships between environmental phenomena and human activities'. It is a single large volume containing both general regional and city maps and many covering special subjects (about 750 maps in all). Some European countries produced national atlases to a high standard considerably earlier; *Atlas de France*, for example, was first produced between 1934 and 1945, and a second loose-leaf edition was published 1953-59. Many of the other national atlases are also loose-leaf and in various stages of completion, and most are produced as co-operative ventures between government bodies and leading geographical and other scientific institutions. *Atlas de Belgique* (1954-), *Atlas over Sverige* (1953-), and *Atlas van Nederland* (1963-) are particularly fine examples. A number of additional ones are individually described by Lock in *Geography and cartography.*

Other regional atlases present great variety, and many are in fact also restricted in scope to certain subjects. Good examples of local ones in Britain are *An atlas of Anglesey* (Anglesey Community Council, 1972); and *Atlas of London and the London region* (Pergamon, 1968), prepared by staff at the London School of Economics and Political Science. Typically, both of these were produced and published as a result of local initiative and expertise, and there are many further examples of such activity. The *North-west regional atlas* (second edition, Department of the Environment, 1976) is an important local atlas produced by a government department. In contrast, there are a number of further examples of regional atlases prepared under the auspices of international agencies or indeed as commercial undertakings, most often covering areas not well served by locally-published maps: *Atlas of physical, economic and social resources of the lower Mekong basin* (United Nations, 1968) is a good example of the former; *Atlas of South-east Asia* (Macmillan, 1964) and the *Times atlas of China* (Times Books, 1974) of the latter. The last mentioned is particularly significant as a contribution to available information on an important area for which it is difficult to obtain up-to-date data. One of the best recent regional atlases, however, is the *Atlas of Europe* (Bartholomew, jointly with Warnes, 1975), which provides useful comparative data in a convenient format on countries already well covered in existing sources. Its strength, apart from the excellent cartography, lies in its collection and juxtaposition of the information, and in its topicality.

The best known series of thematic regional atlases is the economic one produced by the Cartographic Department of the Oxford University Press. The most useful of these are the *Oxford regional economic atlas of Western Europe* (1971) and the *Oxford regional economic atlas of the United States and Canada* (second edition, 1975). These use the same scales and include the same range of topics and style of presentation. There is also a rather older atlas covering the USSR and Eastern Europe (1956) and the *Oxford economic atlas of the world* (fourth edition, 1972). *Business in Britain: a Philip management planning atlas* (Philip, 1975) is designed to 'assist industry, commerce, government and other agencies in making planning decisions by presenting in a clear and concise manner the diverse and complex spatial organisation of economic activity in Britain'. It contains a wealth of economic, social and administrative data. On a much smaller scale there is a *Social atlas of London* (Oxford U P, 1975), using modern techniques of computer graphics to analyse social structure and patterns in a single city.

There are a great many historical atlases; good recently published examples include *Shepherd's historical atlas* (Philip, 1976), and the *New Cambridge modern history atlas* (Cambridge University Press, 1970), in which the maps are also by Philip. The world distribution of landform types and the processes of weathering and erosion which form them are illustrated in an *Atlas of world physical features*, compiled by R E Snead (Wiley, 1972); another physical atlas, *Physikalischer atlas des Heinrich Berghaus*, published in 1845, was probably the first ever thematic atlas, also on a world scale. More specialised is *National atlas of disease mortality in the United Kingdom* by G M Howe (Nelson, 1963). This was the result of a cooperative project between the British Medical Association and the Royal Geographical Society, and was an influential example of the application of cartographic techniques in the analysis of data for social and planning purposes. *An agricultural atlas of England and Wales* by J T Coppock (second edition, Faber, 1976) is another well-established example; this edition contains data up-dated to 1970. The range of possible subjects for presentation in cartographic form is obviously as broad as geography itself, and the trend to increasingly specialised titles is likely to continue; those cited above are merely examples, and constitute only a small sample of the total available.

Seven

SOURCES OF STATISTICS

STATISTICAL INFORMATION is a basic requirement of many disciplines, and geography is no exception. Although traditionally associated with economic studies, statistics are a prominent feature in any regional work as well as in economic geography and other fields of human geography, and, to a lesser extent, in certain aspects of physical geography. The need for statistical information, then, is nearly universal, and the geographer should be aware of the sources which can satisfy this need. Unfortunately, although a wealth of material is available, statistics by their very nature present certain practical problems in use, and it is appropriate to preface this survey of some of the major sources with a brief indication of these difficulties, which are common to the use of statistical data from any source.

Problems in the use of statistics

It is commonplace that statistics are widely abused, and can be manipulated or misinterpreted with disastrous results. In using the various sources described in this chapter, and other statistical material, it is essential therefore that the reasons for this are appreciated. Not all statistics can be accepted without question as accurate. It is not unknown for deliberately misleading figures to be published for political or propaganda purposes, and it is wise to be cautious if there is any reason to suspect such motives. As a routine precaution the authority of the publishing or compiling agency should be noted; for the same reason statistics should never be quoted without citing their source.

Quite apart from the relatively rare occurrence of deliberately inaccurate data, unintentional errors can creep in from several sources. Care must be taken to establish the means used to collect innocently tabulated statistics, and, where appropriate, the size and method of selection if a sample has been used. Many statistical tables incorporate estimated figures to complete a given series in the absence of a properly substantiated return. Again, the purpose of compilation should be noted; statistics collected in response to a particular requirement may be unsuitable for use in other contexts. The best safeguard against errors arising from careless use of statistics in these and similar circumstances is to check carefully any introductory notes, footnotes and other annotations provided. Most statistical works incorporate information on the methods of compilation, scope and area covered, and anomalies such as estimates or a variable data base.

Another common source of trouble concerns the compatibility of statistical data. Units of measurement are not always consistent, even within a single table, in the case of international statistical sources assembled from returns from a variety of agencies. More frequently this problem arises from the need to compare data from different sources. In such cases it is necessary to convert statistics into comparable units, and conversion tables are available for this purpose; *Geographical conversion tables* by D H K Amiran and A P Schick (International Geographical Union, 1961), is an appropriate example devised for the use of geographers. Other problems of compatibility arise where a series of figures include some which are based on slightly varying criteria. Usually this is indicated, as stated above, in a footnote or by some similar device.

Frequently it is important to ascertain the most recent statistics on a particular topic, and this may involve a protracted search. Generally speaking the most readily available and most convenient statistics are found in comprehensive compilations which are only digests of data previously published elsewhere, perhaps in more obscure publications; this is one of the drawbacks of works like, for example, the *Geographical digest*, which is handy for many purposes but not so reliable if really current information is important. Secondary sources such as this have a further disadvantage in that the information contained is often less detailed than in its original format, again possibly necessitating an approach via the primary source. On the other hand, the statistics summarised and edited in ready reference compilations can be most useful for comparative purposes, incorporating conversions into common units and employing whenever possible standardised headings and subdivisions. A good recent example of a very handy source illustrating these features is *The world in figures*, published by the *Economist* in 1976. This brings together selected data from widely scattered original sources on over two hundred countries, and also includes some summary tables and maps.

Many more statistics are collected than actually appear in print, accessible for use. There are of course several reasons for this, one of which

Ports (goods traffic, 000 tonnes, 1973) San José: loaded 135, unloaded 854; Santo Tomás de Castilla: loaded 190, unloaded 450; Puerto Barrios: loaded 362, unloaded 87; Champerico: loaded 71, unloaded 78
Airport La Aurora (Guatemala City)

Durable equipment (at end-year)	000	no per 1 000 people	no per km of road
Radio sets (1974)	260*	46*	
Television sets (1974)	105*	18*	
Passenger cars (1972)	54	10	4.0
Commercial vehicles (1972)	37	7	2.8

Production

Gross domestic product 1974: Q 3 097 mn = $ 3 097 mn = £ 1 324 mn
Growth in real terms: 1960-70 5.5 %pa, 1970-74 6.3 %pa

Production indices (1970 = 100)	1960	1970	1974	Growth %pa 1960-70	1970-74
Agricultural	72	100	121	3.3	4.9
Industrial	58	100	124ᵃ	5.6	7.4ᵇ
Main products (000 t)					
Agriculture					
Maize	506	719	613*	3.6	−3.9*
Sugar, raw value	73	171	325*	8.9	17.4*
Bananas	244	487	450*	7.2	−2.0*
Coffee	99	133	138*	3.0	0.9*
Cotton	21	56*	116*	10.3*	20.0*
Tobacco	1.4	3.2	5.5*	8.6	14.5*
Milk	144*	262	300	6.2*	3.4
Beef and veal	34*	55	62*	4.9*	3.0*
Timber (000 m³)	7 000*	7 003*	7 017ᵃ	0.0*	0.1*ᵇ
Other					
Petroleum products	—	724	922*ᵃ	na	8.4*ᵇ
Electricity (mn kW h)	281	780	910*ᶜ	10.7	8.0*ᵈ
Antimony ore (Sb content)	—	1.30	0.96ᵃ	na	9.6ᵇ
Tungsten conc (oxide content)	—	0.05	0.20ᵃ	na	58.5ᵇ
Beer (000 hl)	166	299	421ᵃ	6.1	12.1ᵇ
Cigarettes (mn units)	1 889	2 986	3 016ᵃ	4.7	0.3ᵇ
Cement	117	251	316ᵃ	7.9	8.0ᵇ

[a]1973 [b]1970-73 [c]1972 [d]1970-72

Transport traffic	1960	1970	1974	Growth %pa 1960-70	1970-74
Passenger-kilometres (mn)					
Air	30	104	100	13.2	−1.0
Cargo: tonne-kilometres (mn)					
Rail	270	106	na	−8.9	na
Air	2.2	6.8	4.8	11.9	−8.3
Sea: tonnes (mn)					
Goods loaded	0.32	0.53	1.10ᵃ	5.0	27.6ᵇ
Goods unloaded	0.59	1.30	1.87ᵃ	8.2	12.8ᵇ

[a]1973 [b]1970-73

Finance and trade

Consumer price index (1970 = 100) 1974 132.7; growth 1970-74 7.3 %pa
Money stock (end-year, Q mn) 1974 314; growth 1970-74 14.8 %pa
Budget (1974) Revenue: Q 282 mn = $ 282 mn = £ 121 mn
Expenditure: Q 328 mn = $ 328 mn = £ 140 mn

Balance of payments ($ mn)	1970	1971	1972	1973	1974
Balance of goods (fob)	+30	−3	+41	+51	−49
Balance of services	−56	−71	−83	−85	−105
Balance of transfers	+17	+25	+30	+43	+55
Current balance	*−8*	*−49*	*−11*	*+8*	*−99*
Long-term capital flow	+55	+46	+33	+60	+70
International reserves (end-year, $ mn)	78	93	135	212	202

External trade (1974) Imports: Q 700 mn = $ 700 mn = £ 299 mn
Exports: Q 586 mn = $ 586 mn = £ 251 mn

Main imports (1972)	% of total	*Main exports* (1974)	% of total
Chemicals	20	Coffee	28
Machinery, non-electric	13	Cotton	11
Textile yarns and fabrics	9	Sugar	11
Food	7	Bananas	6
Motor vehicles	7	Beef and veal	4
Main sources (1974)		*Main destinations* (1974)	
United States	32	United States	33
Venezuela	12	El Salvador	11
El Salvador	10	West Germany	11
Japan	9	Nicaragua	7

Guyana

Location North east of South America
With a coastline on the Atlantic Ocean, Brazil is south, Venezuela west and Surinam east
Land Area 215 000 km² = 83 000 km²
Climate Tropical inland, sub-tropical on coast
Weather at Georgetown, 2 m altitude
Temperature: hottest months Sept, Oct 24-31 °C, coldest Jan, Feb 23-29 °C
Rainfall (av monthly): driest month Oct 76 mm, wettest June 302 mm
Time 3¾ hours behind GMT (winter time, 3 hours behind)
Measures UK (imperial) system: also metric system and Rhynland acre = 0.426 hectare = 1.052 acres
Monetary unit Guyanese dollar (G $) = 100 cents
Rate of exchange (1974 av): par G $ 5.2114 = £ 1, free G $ 2.228 = $ 1
The link with the pound was changed from October 9, 1975 to a link with the United States dollar at G $ 2.55 = $ 1

Summary

Political Co-operative republic, which became independent May 26, 1966 formerly the UK colony of British Guiana. Member of UN, Commonwealth, Caricom and an EEC ACP state
Economic Sugar is the main export, and the agricultural sector makes up 17 % of gross domestic product; mining is also important, making up 13 % of gross domestic product, with bauxite and alumina accounting for one-third of exports. Industry, especially textiles, is to be developed

People, resources and equipment

Population 1960 565 000, 1970 718 000, 1974 774 000*
Growth: 1960-70 2.4 %pa, 1970-74 1.9* %pa
Density (1974): 4* people per km²
Vital statistics (rate per 1 000 people, 1970) births 34.3, deaths 6.6
Cities (population in 000, 1970) Georgetown (capital) 168, Linden 30*, New Amsterdam 18*, Corriverton 17*, Rose Hall 8*
Race (1970) Asian Indian 51 %, African 31 %, Mixed 12 %, American Indian 4 %, Portuguese 1 %, Chinese 1 %
Language English; Creole and American Indian languages are also used
Religion (1960) Hindu 33 %, Anglican 19 %, Roman Catholic 15 %, Moslem 9 %, Methodist 4 %
Education (1970/71) Pupils 192 525, teachers 6 965
Labour force (1970) 229 000; in agriculture 75 000 (32 %)
Personnel (1972) Physicians: 191, 1 per 3 910 people
Standard of living
National income per person (1974): G $ 910** = $ 410** = £ 175**
Consumption per person (1973): energy 950 kg coal equivalent, electricity (production) 503 kW h, newsprint 2.2 kg
Newspapers (1972): number 4; circulation 41 000, 55 per 1 000 people
Telephones (Dec 1973): 17 500, 23 per 1 000 people
Livestock (000, 1973/74) Cattle 265*, sheep 108*, pigs 106*, chickens 8 850*
Electrical capacity (public only, 1971) 90 megawatts
Hospital beds (1972) 3 969, 1 per 190 people
Roads (1973) 2 910 km = 1 810 mi, density 0.01 km per km²
Railways (1973) 29 km = 18 mi, density 0.0001 km per km²
Ships (registered, 1975) 63, total of 16 828 gross tons
Ports Georgetown, New Amsterdam
Airports Timehri (37 km from Georgetown); also 21 other airports with scheduled flights
Durable equipment (at end-year)
Radio sets (1972): 100 000, 133 per 1 000 people
Passenger cars (1973): 23 000, 30 per 1 000 people, 8 per km of road
Commercial vehicles (1973): 10 000, 13 per 1 000 people, 3 per km of road

Production

Gross domestic product 1973: G $ 643 mn = $ 302 mn = £ 123 mn
1974 est: G $ 780** mn = $ 350** mn = £ 150** mn
Structure of gross domestic product (1973) *By origin* Agriculture 17 %, mining and quarrying 13 %, manufacturing 10 %, other 60 %

Production index (1970 = 100)	1960	1970	1974	Growth %pa 1960-70	1970-74
Agricultural	90	100	117	1.1	4.1

Figure 15: Sample page from 'The world in figures'

is the unhelpful one that the information may be confidential. In many cases, however, publication does not take place simply because it cannot economically be justified, but the data may nevertheless be available on request. In these circumstances, it is worth tracing the potential producers of required statistics and making inquiries. This is perhaps most significant for the geographer in the context of local statistics, which are very difficult to ascertain from published sources, although records may be maintained by local government offices, chambers of commerce, and a wide variety of associations and societies. On a national scale, one of the more important statistical functions of the *Statesman's yearbook* is its practice of listing the central agency responsible for the collection of statistics in each country. (Also included are details of major published statistical sources, and of course actual statistics, though the latter are subject to the same reservations as those already expressed about the *Geographical digest*.) The major national statistical serials for each country are also listed in the final chapter of *How to find out about statistics*, G A Burrington (Pergamon, 1972); most of this slim volume, however, is concerned with broader aspects of the nature and use of statistics.

Statistical bibliographies and libraries

While a great many statistical works are published and accessible, it is only possible here to describe a small selection of those most likely to be required for geographical purposes. As with other categories of literature included in this book, use should be made of the bibliographies, guides, and library collections through which may more titles can be traced.

Unfortunately there is no single international bibliography in which all or even a large selection of the more important statistical collections and series are listed. Perhaps the nearest approach to such a general guide is *Sources of statistics* by J M Harvey (second edition, Bingley; Hamden (Conn) Linnet, 1971). While this is very good on British statistics, and includes a selection of American and international sources, it by no means satisfies comprehensively the requirements of geographical study involving other parts of the world. A shorter check-list prepared by a joint working party of the Royal Statistical Society and the Library Association is a useful guide to sources which can be expected to be readily available in public libraries; this is published in two parts as *Recommended basic United Kingdom statistical sources for community use* (third edition, 1975) and *Recommended basic statistical sources: international* (1975). Current developments in the publication of official British statistics are reported in the Central Statistical Office's *Statistical news: developments in British official statistics* (HMSO, quarterly, 1968-) which can be used to supplement *Guide to official statistics*, published in 1976 by the government Central Statistical Office. This is an annotated list, arranged under subjects and with a keyword index, which also provides appropriate departmental telephone numbers to facilitate the search for further information, including unpublished data. A fuller and more critical discussion of the main UK sources is to be found in two books by B Edwards published by Heinemann, entitled *Sources of social statistics* (1974) and *Sources of economic and business statistics* (1972). Very full descriptions, including details of the methods of collection (obviously essential to any serious use of statistics) are also found in *Reviews of United Kingdom statistical sources* edited by W F Maunder, published by Heinemann for the Royal Statistical Society and the Social Science Research Council, 1974-5. The four volumes of this set each contain two extensive reviews by specialists on major subject areas.

Similar information for the United States is contained in *Statistical reporter* (monthly, 1966-) issued by the Bureau of the Budget, which updates information published in an earlier summary, *Statistical services of the United States government* (revised edition, 1968). Also useful is an unofficial guide, *Guide to US government statistics* by J L Andriot (third edition, Documents Index, 1961). This lists nearly two thousand statistical series, arranged by the responsible government department, and a subject index is provided. Of broader scope, including both unofficial sources and non-American publications, is *Statistics sources: a subject guide to data on industrial, business, social, educational, financial and other topics for the United States and internationally.* Edited by P Wasserman and J Bernero, and published by Gale Research (fifth edition, 1977) this contains over twenty thousand citations of statistical data on nearly twelve thousand subjects, with a select list of key sources.

Bibliographies of statistics which are genuinely international in scope are, unfortunately for the geographer, much rarer. Statistical series and other compilations published by international organisations such as the United Nations and its agencies can be traced through the general catalogues of these bodies, used to supplement the rather dated *List of statistical series collected by international organisations* (United Nations Statistical Office, 1955). Other early and also dated sources include some bibliographies issued by the Library of Congress; for example, *Statistical yearbooks: an annotated bibliography of the general statistical yearbooks of the major political subdivisions of the world* (Government Printing Office, 1953), and a companion volume on *Statistical bulletins . . .* (1954).

A selective current bibliography which represents publications received on exchange by the United States Bureau of the Census is *Foreign statistical publications* (quarterly, 1947-). Another similarly limited bibliography of American origin and excluding American publications is *Foreign statistical documents*, edited by J Ball (Stanford University Press, 1967).

There are in addition to the above a number of more obviously restricted bibliographies, limited either by subject or by geographical area. An example of the former which is most useful to geographers is *International population census bibliography*, prepared by the Population Research Center, University of Texas. Originally published (1965-7) in six volumes covering the continents, there is a further supplementary volume bringing the original set up to date to 1968. Although concerned principally with vital statistics, where other data was issued as an integral part of a census return, this is cited. While the bibliography does not claim comprehensiveness the world-wide coverage makes this a most valuable source of information on population statistics. Two further examples covering other fields are compiled by GATT, the General Agreement on Tariffs and Trade; *Compendium of sources: international trade statistics*, and *Compendium of sources: basic commodity statistics*, both published in 1967, and listing sources both national and international. Examples of area limitation are four volumes by J Harvey and published by CBD Research. The first of these, *Statistics Europe: sources for social, economic and market research* was originally published in 1968 but now, in its third edition (1976) is the most up-to-date in the series. Major sources are listed on a country-by-country basis, with full bibliographical details and brief notes on content; a sample page is illustrated in figure sixteen. The other volumes, in a uniform format, are *Statistics Africa* (1970); *Statistics America* (1973), which covers north, central and south America; and *Statistics Asia and Australasia* (1974). These four together comprise a near-comprehensive guide to all the main sources to which the geographer is likely to require access.

From the above it will be appreciated that statistical literature is a complex field in its own right (comparable in this sense with maps), and while a selection of basic materials is available in many libraries, comprehensive or very extensive coverage necessitates dependence on specialist collections. The most comprehensive collection in the UK of current statistical material is the Statistics and Market Intelligence Library of the Department of Trade and Industry, some of whose bibliographic publications have already been noted. A second collection of some importance

¶ C - External trade

816 Maandstatistiek van de buitenlandse handel per goederensoort [Monthly statistical bulletin of foreign trade by commodities] (Centraal Bureau voor de Statistiek).

Staatsuitgeverij, Christoffel Plantijnstraat, Den Haag; or from Dutch booksellers.

1946- Fl 7.50; or Fl 75.00 yr.

Detailed monthly and cumulative figures of imports and exports by commodities (BTN) sub-divided by countries; exchange trade with the Belgo-Luxembourgeoise Union; imports and exports from and to the Belgo-Luxembourgeoise Union; and index numbers of foreign trade.

Time factor: each issue carries statistics for that month and cumulations for the year to date, and is published four to six weeks after the end of the month. The December issue contains the final annual figures as well as those for December.

§ Nl; list of contents and table headings also in En.

Note: Naamlijsten voor de statistiek van de buitenlandse handel, 1 januari 1974 [List of goods for the statistics of foreign trade] is also available.

817 Maandstatistiek van de buitenlandse handel per land [Monthly statistical bulletin of foreign trade by countries] (Centraal Bureau voor de Statistiek).

Staatsuitgeverij, Christoffel Plantijnstraat, Den Haag; or from Dutch booksellers.

1946- Fl 4.50; or Fl 45.00 yr.

Cumulative statistics of imports and exports (parcel post included) according to SITC groups and according to chapters of the BTN for a number of countries. Also totals of storage in and removal from bonded warehouses by countries.

Time factor: each issue has statistics for that month and cumulations for the year to date, and is published about two months after the end of the period covered. The December issue has the final annual figures as well as those for December.

§ Nl; list of contents and table headings also in En.

¶ D - Internal distribution and service trades

818 Maandstatistiek van de binnenlandse handel [Monthly bulletin of distribution statistics] (Centraal Bureau voor de Statistiek).

Staatsuitgeverij, Christoffel Plantijnstraat, Den Haag; or from Dutch booksellers.

1953- Fl 5.25; or Fl 52.50 yr.

Contains monthly figures on turnover and stocks for a large number of branches in wholesale, retail and servicing trades; on credits granted and outstanding; and on buyers' credits in the retail trade. Monthly data on the quantities of a number of food and tobacco products and durable consumer goods that become available for domestic consumption. Monthly consumer price index numbers; wholesale price index numbers (producers', import and export prices) for a large number of groups of industry, often with a very detailed specification according to groups of commodities; wholesale prices of a number of important durable and non-durable goods at the various world markets and in the Netherlands.

Time factor: each issue has data for several complete years and several months up to about four months prior to the date of the issue.

§ Nl; list of contents and table headings also in En.

Figure 16: Sample page from 'Statistics Europe'

is housed in the University of Warwick library. In addition, a few of the larger public libraries have established above average collections in response to local commercial needs, and these may prove useful for certain categories of material.

International sources–the world

The most important publisher of statistics which are international in scope is the Statistical Office of the United Nations, and the best known series is the *Statistical yearbook.* Published annually since 1947 by the United Nations this is a successor to an earlier work of the same name previously published (1926-) by the League of Nations. More than 250 countries and territories are featured in the tables and in subject scope the topics covered include population, agriculture, industries, trade, finance, and education. There is also a more current publication, *Monthly bulletin of statistics* (1947-) which up-dates the information in the yearbook, which is inevitably dated. A second annual publication from the same source and in a uniform format is the *Demographic yearbook* (1948-), containing statistics on population, marriage, births, deaths and migration. Both yearbooks are bilingual (English and French) and in both it is customary for the tables to cover a span of several years for comparative purposes. A third important example from the United Nations is the *Yearbook of international trade statistics* (1950-) which includes some summary tables but consists mainly of tables for about 150 individual countries, arranged wherever possible according to the standard international trade classification. Again there is a more frequently published source, the quarterly *Commodity trade statistics* (1952-) which provides up-to-date information, arranged by country. The same data is reorganised and published as *World trade annual* (Walker and Co, 1963-), the five volumes of which are arranged under commodity types, subdivided by countries of origin and destination.

An important producer of statistics relating to agriculture and allied activities is the Food and Agriculture Organisation of the United Nations. The *Yearbook of food and agricultural statistics* first appeared in 1947 and was published for a number of years in two parts, entitled 'Production' and 'Trade'. This publication was succeeded by the *Production yearbook* and the *Trade yearbook.* These include detailed data on specific crops, livestock, machinery, fertilizers, land use, wages, and so on. FAO also produce an annual review, *The state of food and agriculture* (1947-), which is a textual summary supported by statistical tables, together with articles on topical subjects. Another very useful summary volume is *World crop statistics: area, production, and yield 1948-1964,* which is an historical digest of material on crops included in the annual series in the *Production yearbook.* As is typical of most UN annual statistical compilations, these large volumes are backed up by a *Monthly bulletin of agricultural economics and statistics* (1952-), providing more current data. The FAO is responsible also for a range of specialised statistical sources, notably the *Yearbook of forest products statistics, Yearbook of fishery statistics*, and the Commodity bulletin series, all of which commenced publication in 1947. The last mentioned is a series of monograph reviews, published irregularly, each devoted to a particular product. Information on individual crops and products is also contained in the various statistical services of the Commonwealth Secretariat, formerly the Commonwealth Economic Committee. These include annual reviews on *Dairy produce, Grain crops, Fruit, Meat, Vegetable oils and oilseeds*, and *Plantation crops*, as well as several monthly and quarterly series.

Industrial activity is, by comparison with agriculture, rather less well documented. The UN Statistical Office produces *World energy supplies* as series J of its statistical papers. The first volume, published in 1952, gave statistics for production, trade, and consumption of commercial sources of energy for the years 1929-50; more recently, publication has become regular (annual). In series P is *The growth of world industry, 1938-61,* issued in two volumes of 'National tables', and 'International analyses and tables' respectively. Hopefully this source too will assume regular and frequent publication. Most other sources of general industrial statistics are more limited by area, and are discussed below, but there are several sources for specialised industrial topics worth noting here. *Oil: world statistics*, an annual volume issued by the Institute of Petroleum, includes data on production, consumption, trade, etc, arranged by countries. The building industry is covered by the *Yearbook of construction statistics* (UN Statistical Office, 1963/72-) and there is an international source of quantitative information on the motor industry in *World automotive market* (Johnston International Publishing Company, 1966-). Other comprehensive annual series on major industries include *Statistical summary of the mineral industry* (Institute of Geological Sciences, 1913/20-), *Wood pulp and fiber statistics* (American Paper Institute, 1937-), and *Industrial fibres* (Commonwealth Secretariat, 1966-). Monthly series of course provide much more current data, but are not available in all fields; two examples well worth a mention are *World metal statistics* (World Bureau of Metal Statistics, 1948-), with information on metals of all kinds, and *Textile organon* (Textile Economics Bureau, 1930-), which includes figures for wool and for man made fibres.

Labour statistics, including information on employment, unemployment, wages, hours of work, productivity, and other topics for more than 170 countries are contained in the *Yearbook of labour statistics* (1941-), published by the International Labour Office. This is kept up-to-date by the quarterly *Bulletin of labour statistics* (1965-). Information on education, media of communication, and other cultural matters is available in the *Statistical yearbook* published by Unesco since 1963. Transport and tourism are covered in *International railway statistics* (International Union of Railways, 1925-), *World air transport statistics* (International Air Transport Association, 1957-), *World road statistics* (International Road Federation, 1965-), and *International travel statistics* (International Union of Official Travel Organisations, 1947-).

All of the above sources reflect interests in economic and social fields, the area of human geography, and this is as far as most of the general guides to statistics go. The most important published statistical data in physical geography is climatic, and several works may be noted which provide details on a global scale. The World Meteorological Organisation has published a *Catalogue of meteorological data for research* (1965-, loose-leaf), which gives statistical and bibliographical information for eighty-eight member nations. The same body has sponsored the continuing publication of a series of reports entitled *World weather records*, earlier volumes being published by the Smithsonian Institution. The most recent volume in the series covers the period 1951-60 in six volumes. Selective information for a variety of stations all over the world is available in *Tables of temperature, relative humidity and precipitation for the world*, published by the UK Meteorological Office, 1968, also in six volumes. A review of some of the problems in tracing meteorological information, together with proposals for improving its availability, is found in a report from the WMO, *World weather watch: collection, storage and retrieval of meteorological data* (1969).

International sources–major groupings and regions

In addition to the several large yearbooks and other publications already cited covering the world as a whole the United Nations, through a number of regional commissions, publishes data for more restricted areas. Economic commissions for Europe, Africa, Latin America, and Asia and the Far East each produce an annual review, variously entitled *Economic survey of . . .* or *Economic bulletin for . . .* These date from 1947, 1962, 1948, and 1947 respectively. Significantly, these annual publications cover between them a great many of the developing countries, on which statistical information is generally less readily accessible from national yearbooks and similar sources. While these annual compilations are in many ways the most convenient and most useful to geographers of the statistical publications of ECE, ECA, ECLA, and ECAFE, mention should also be made of the more detailed series produced by all four bodies: for example the *Annual bulletin of gas statistics for Europe* (1957-) and parallel series on coal and electricity, *Foreign trade statistics of Africa* (1962-), or *Timber bulletin for Europe* (1948-). A variety of other similar and equally important series published by these commissions may be traced in the bibliographies of statistics already described.

A further organisation which is an important publisher of international statistical data is the Organisation for Economic Cooperation and Development (OECD), which comprises over twenty member states. These include the UK, USA, Canada, and Japan, as well as many European nations. The principal general series from OECD is a monthly publication called *Main economic indicators* (1965-) which superseded *Bulletin of general statistics.* This documents the statistical facts underlying recent changes and developments in the economy of each country. Supplements are also published covering consumer prices and industrial production, as well as less frequent volumes of historical statistics.

Apart from this there are numerous more specialised statistical works on particular topics. *Statistics of foreign trade* (1950-) is published in three series: *Overall trade by countries* (series A, quarterly); *Trade by commodities: analytical abstracts* (series B, quarterly, each issue being in six separate parts); and *Trade by commodities: market summaries* (series C, twice yearly). OECD does not produce a separate serial covering industrial production, although there is a useful summary volume, *Industrial statistics, 1900-1962* (1966). A similar monograph publication which provides data on main agricultural products, but covering a shorter period, is *Agricultural statistics, 1955-1968* (1970). In addition there are certain detailed series covering specific industries, such as *Oil statistics* (1961-), *Timber market in the OECD countries* (1953-), *Textile industry in OECD countries* (1953-), *Review of fisheries in OECD member countries* (1967-), and *Tourism in OECD member countries* (1962-). Other industries, including pulp and paper, cement, iron and steel, and non-ferrous metals are the subject of similar publications. Concerned with more than just a single industry are the annual *Statistics of energy* (1950/7-) and the biennial *Labour force statistics* (1955/66-). Less detailed selective statistics are available in annual *Economic surveys* of each country comparable to the regional collections produced by the UN economic commissions.

Among important trading blocs EFTA is not a prolific publisher, although the annual *EFTA trade* (1959/63-) should be noted, while the European communities produce a wealth of statistical material. The three communities, European Economic Community (EEC), European Coal and Steel Community (ECSC) and European Atomic Energy Community (EURATOM) are served by a common statistical office, which publishes the annual *Basic statistics of the community* (1960-) and a more frequent (monthly) *General statistical bulletin* (1961-). The anual volume also includes figures for selected countries outside the community for comparative purposes. Foreign trade statistics of members and associated states are listed in *Foreign trade: monthly statistics, Foreign trade: analytical tables* (quarterly) and *Overseas associates: foreign trade statistics* (monthly). Agricultural and industrial production respectively are covered in *Agricultural statistics* (1961-) and *Industrial statistics* (both quarterly and annual, 1960-); the publishing pattern of the latter is repeated in a further series of *Energy statistics* (1953-). EEC economic data is available monthly in a rather different form in *Graphs and notes on the economic situation in the community* (1959-). There is also a useful bi-monthly series of *Social statistics*, in which each issue deals with a particular topic or problem. Some subjects recur at more or less annual intervals, or even more frequently. The European Communities' Statistical Office also produces an annual volume of *Regional statistics* (1971-), containing data on a wide range of topics.

Other smaller European political groups which combine to produce international statistics are Benelux, with *Bulletin trimestriel économique et statistique* (1954-) which contains a statistical section, and the Nordic Council (Nordisk Råd) with *Yearbook of Nordic statistics* (1962-) which provides a convenient summary, in English, of data from the national statistical publications of the five member states. In Africa there is the Organisation Commune Africaine et Malgache (OCAM) with an annual *Compendium des statistiques du commerce extérieur* (1965-) and a quarterly *Bulletin statistique de l'OCAM.* The Pan American Union (PAU) produces every two years *America en cifras* (1960-) and a monthly *Boletin estadistico* (1965-), covering north, south and central America and certain Caribbean countries. A further source on Latin America is *Statistical abstract of Latin America* (1956-), published by the Center of Latin American Studies of the University of California, Los Angeles.

The yearbooks and statistical serials discussed in both this section and in the preceding paragraphs collectively provide information on most of the countries of the world. Their general breadth, however, is a disadvantage where more detail is required for intensive regional study, and to this end it may be necessary to consult more localised sources. In many cases the statistics contained in the foregoing works are reproduced from alternative sources produced locally, usually with some sacrifice of detail. While the originals may be difficult to locate and certainly are too numerous to be listed here, some general comment is appropriate. Firstly, almost every country has some agency responsible for the collection and publication of statistical data, and as already noted the *Statesman's yearbook* is a readily accessible source of information, listing both basic national statistical publications (most characteristically yearbooks) and the bodies responsible for their compilation. This should be supplemented by the rather fuller bibliographies already detailed. Secondly, there are numerous journals produced by leading banks in a great many countries which frequently contain valuable statistical data. These are especially useful for information on the developing countries, where official statistics can be difficult to trace.

British and American statistics

As examples of the complex network of statistical works which are available covering many of the industrialised nations, we now take a brief look at some of the major British and American sources. In both countries there is a profusion of material, and several of the bibliographies noted previously concentrate on listing this in far greater depth than is possible here, where only the main general sources are described. For more specific information on industry, agriculture, trade, etc, these further specialised works should be used.

The official annual compendium of statistical information in the UK is the *Annual abstract of statistics* (HMSO, 1946-), which is a continuation of the former *Statistical abstract of the United Kingdom*, issued by the Board of Trade and providing figures on a wide range of topics back to 1840. This is backed up by the *Monthly digest of statistics* (1946-) and two newer journals of more limited scope which however cover between them a considerable field, namely *Economic trends* (1953-) and *Social trends* (1970-), all published by HMSO.

Two useful and concise handbooks, both published by Penguin Books, are *Facts in focus*, compiled by the Central Statistical Office (1972) and *Britain in figures: a handbook of social statistics* by A F

Sillitoe (1971). The latter presents the data graphically, with a commentary. *Statistical abstract of the United States* (1878-) is comparable in scope with the British *Annual abstract*, and both include climatic data as well as the more usual series on economic and social subjects. Economic statistics are up-dated monthly by *Economic indicators* (1948-), issued by the Council of Economic Advisors, and the *Handbook of basic economic statistics* (Economic Statistics Bureau, Washington, 1947-), also monthly, and slightly more up-to-date than the official series.

Although the main annual series in both countries date from the nineteenth century, providing a long run of historical statistical information, it is worth noting two monograph publications which summarise these in a convenient format. *Historical statistics of the United States: colonial times to 1957* is an official production of the US Bureau of the Census, published in 1960. A supplement (1965) takes the work up to 1962. For the United Kingdom there is *Abstract of British historical statistics*, by B R Mitchell and P Deane (Cambridge UP, 1962). Both works cover a wide range of topics of interest to the geographer, and the notes on sources provide useful clues for further research. *Second abstract of British historical statistics*, by B R Mitchell and H G Jones (CUP, 1971) continues where possible the series contained in the earlier volume by Mitchell and Deane from 1938 to 1965/6.

Regional statistics are also of great geographical relevance, and there has been a considerable improvement in their provision in recent years. The Central Statistical Office produces an annual *Abstract of regional statistics* (1965-), which brings together statistics which are available on a regional basis, based largely on the Standard Regions. Scotland, Wales, and Northern Ireland are represented in the *Digest of Scottish statistics* (1953-), *Digest of Welsh statistics* (1954-), and *Digest of statistics Northern Ireland* (1954-); the first and last mentioned of these are published twice yearly, and the volume for Wales is annual. For the United States there is the *County and city data book*, which appears somewhat irregularly as a supplement to *Statistical abstract of the US.* In addition, most individual states issue their own series of statistical yearbooks: a list of these is given in Harvey's *Statistics America*, cited above.

Eight

GOVERNMENT AND INTERNATIONAL ORGANISATIONS' PUBLICATIONS

GOVERNMENTS ARE among the world's most prolific publishers, producing not only a vast quantity of original factual material (reports and statistics) but also analytical studies which have a bearing on scholarship in many disciplines. The statistical output of both governments and official international bodies featured largely in the previous chapter, but this is but a small part of the total. A bibliography of American government documents published since 1945 identifies 3500 items of special interest to geographers, but warns that 'this listing should be viewed as a small sample useful in introducing the reader to the vast research output of the Federal government'.

The work in question, *US government publications for research and teaching in geography and related social and natural sciences* by C L Vinge and A G Vinge (Littlefield, Adams, 1967) is a revised version of a shorter list originally published by the National Council for Geographic Education, 1962. A quick glance at the table of contents giving a list of departments represented in the guide is sufficient to give a striking impression of the great variety of topics involved; these range from the armed services to the Post Office, and from the Civil Rights Commission to the Atomic Energy Commission, and include also a number with names more suggestive of their relevance to geography—the Weather Bureau, the Geological Survey, the Geographic Names Board.

In view of this extreme diversity, which is typical of other governments in addition to the United States', and in view of the impossibility of meaningful selection from so many documents within the compass of this book, attention here is confined to those bibliographies and guides through which the original publications may be traced. In passing it may be noted that government publications themselves include a large number of bibliographies, and pages 241-263 of the above mentioned title are devoted to a list of some examples produced by departments of the US government. There is unfortunately no single source listing bibliographies and guides to government publications of all countries, and in this chapter it is possible only to describe British and American examples; the nearest approach to a comprehensive list is *A study of current bibliographies of national official publications* by J Meyriat (Unesco, 1958), which is now rather out of date.

British government publications

An introduction to British government publications by J G Ollé (second edition, Association of Assistant Librarians, 1973), is a most useful guide to the complex patterns of government publishing in the UK. This is a concise account written primarily for student librarians, providing an elementary introduction to the subject, indicating and defining the various categories of publication, and citing examples. A more substantial, though less readable, volume containing additional factual information is *British official publications*, by J E Pemberton (second edition, Pergamon, 1973). Again the various categories are described, their usefulness and content explained, and their bibliographic organisation through catalogues and indexes is clearly demonstrated. The book makes effective use of sample pages from various typical examples of government documents. In addition to these two major guides there is a short official sourcebook issued by the Treasury and entitled simply *Official publications* (1958, reprinted 1963) which although now dated does provide an authoritative statement on the principles behind the definitions of each of the several categories.

Briefly, British official publications may be classified into two categories. Parliamentary papers are basically publications resulting directly from the activities of parliamentary procedure: bills, acts, and the transcripts of the debates of both Houses of Parliament are the most obvious examples, but also included are reports of certain committees relating to existing or proposed legislation. Non-parliamentary publications are the products of ministries and government departments issued for more general use, and not just in connection with the legislative process. These statements, however, are generalisations, and it is not difficult to find exceptions, especially as government publishing policy has changed somewhat over the years. Some of the available indexes and bibliographies cover both categories, and some are confined to only one group.

Each parliamentary session produces a collection of papers and bills which are indexed in a *Sessional index*. These indexes cumulate into decennial indexes which in turn are consolidated into fifty year sets, the most recent of which covers the period 1900-1949. Reports of debates are contained in *Hansard*, which is the abbreviated name by which the official record is usually known. *Hansard* dates from 1803, and since the fifth series was begun in 1909 proceedings in the House of Lords have been reported separately. The two editions, for the two Houses of Parliament, are currently both published daily, weekly, and finally in bound volumes based on complete sessions. There are weekly and sessional indexes.

Among several non-official indexes and guides to parliamentary papers are a number by P and G Ford, whose *A guide to parliamentary papers* (third edition, Blackwell, 1972) is a concise summary of the field. Three volumes of *A breviate of parliamentary papers* (Blackwell, 1951-61) cover the years 1900-1916, 1917-1939, and 1940-1954, providing abstracts of reports of commissions and committees. Some of the material is in fact non-parliamentary. For the earlier period the same authors have produced a *Select list of British parliamentary papers, 1833-1899* (Blackwell, 1953) which lists important items under subject, without annotation or abstracts, and *Hansard's catalogue and breviate of parliamentary papers 1696-1834* (Blackwell, 1953). *Guide to the records of Parliament* by M F Bond (HMSO, 1971), deals with both manuscript and printed records preserved at Westminster dating from the middle ages to the parliamentary session of 1969/70, incorporating detailed explanatory notes and references.

Non-parliamentary publications may be traced in a number of ways, one of the most useful tools being a series of *Sectional lists*, each of which covers a particular department, group of departments, or subject. Each catalogue lists only publications which are currently in print, and is revised periodically. There are currently about forty of these lists, and the arrangement is by department and/or series, which does not facilitate searching by subject, and it is not always easy to guess which is the appropriate list for a given topic. The format is illustrated in figure seventeen, which shows a page from the *Sectional list of Meteorological Office publications.*

A current bibliographical service incorporating both categories of publication, and the earliest notification of the appearance of government documents, is the *Daily list of government publications from Her Majesty's Stationery Office.* Also indicated in this list are publications sold but not published by HMSO; this is the situation in the case of some publications from international organisations, with the Stationery Office acting as agents. The daily list cumulates into a monthly catalogue, to which an index is provided. For retrospective bibliographical checking there is an annual list, entitled *Catalogue of government publications*

Charts of Marine Meteorology, Sea-Surface Currents and Ice—*contd.*

†Indian Ocean:

Monthly Charts of Dew-point Temperatures over the Indian Ocean. (Met.O. 812. 1969) (11 400101 4)

‡Pacific Ocean:

Monthly Meteorological Charts of the Eastern Pacific. (Met.O. 518. 2nd edition 1968) Prepared in the Marine Division of the Meteorological Office. (40–135–0–68)

Quarterly Surface Current Charts of Western North Pacific Ocean westward of longitude 160°W, with monthly chartlets of the China Seas. (Met.O. 485. 2nd edition 1949, *reprinted* 1970) (11 400094 8)

Quarterly Surface Current Charts of Eastern North Pacific Ocean eastward of longitude 160°W. (Met.O. 655. 1960) (40–162)

†Northern Hemisphere

Supplied in monthly parts with title page for binding:

Monthly Ice Charts 1972. (Met.O. 851. 1973)
Monthly Ice Charts 1973. (Met.O. 867. 1974)
Monthly Ice Charts 1974. (Met.O. 873. 1975)
Monthly Ice Charts 1975. (Met.O. 882. 1976)
Monthly Ice Charts 1976. (Met.O. 898. 1977)
§Monthly Ice Charts 1977. (Met.O. 906. *In preparation*)

‡*Superseded by* Admiralty Routeing Charts. *Available at all Admiralty Chart Agents in the U.K., not HMSO*

WEATHER OVER THE OCEANS AND COASTAL REGIONS

Weather in the China Seas and in the western parts of the North Pacific Ocean. (Met.O. 404):

Vol. II. Local Information. (Met.O. 404b. 1937, *reprinted* 1945) (40–139–2) (4to) 390g £1·62½p

Vol. III. Aids to Forecasting, including supplement on Single Observer Forecasting. (Met.O. 404c. 1938, *reprinted* 1945) (40–139–3) (4to) 390g £2·25

Weather in Home Waters. (Met.O. 732):

Vol. I. The Northern Seas (Norwegian and Barents Seas and East Atlantic North of 60°N). **Part 1.** (Met.O. 732a. 1964) (40–171–1*) 1·8kg £2·75

The Northern Seas **Part 2**—General Tables (Met.O. 732b. 1965) (40–171–2*) 1·8kg £1·75

Vol. II. The waters around the British Isles and the Baltic. **Part 1** (with corrections) (Met.O. 732c. 1975) (0 11 400284 3*) 1.25 kg £18

§**Part 2.** (Met.O. 732d. 1977. *In preparation*)

§**Vol. III.** The waters around The Azores and off south-west Europe and off north-west Africa. **Parts I and II.** (Met.O. 732e and Met.O. 732f. *In preparation*)

Weather in the Mediterranean. (Met.O. 391):

Vol. I. General Meteorology. (Met.O. 391a. 2nd edition 1962, second impression 1970) (11 400109 X) 1·8kg £4·75

Vol. II. Local Information. (Met.O. 391b. 2nd edition 1964) (40–142–2–63) 675g £2·10

RESEARCHES AND APPLIED METEOROLOGY

Geophysical Memoirs

Vol. XI:

89. **Temperature and Humidity Gradients in the first 100 m. over south-east England.** By A. C. Best, M.Sc., E. Knighting, B.Sc., R. H. Pedlow, B.Sc., and K. Stor-

Figure 17: Sample page from the Meteorological Office 'Sectional list'

Series AB

1 Abortion statistics. legal abortions carried out under the 1967 Abortion act in England and Wales, 1974; vi, 58p.; mostly tables; 30cm [0 11 690635 9] 200g; £1.75

Series DH4

2 Mortality statistics. accidents and violence: review of the Registrar General on deaths attributed to accidental and violent causes in England and Wales, 1975; viii, 38p.of tables; 30cm [0 11 690639 1] 150g; £1.25

Series PP2

7 Population projections. population projections by sex, age and marital status for United Kingdom and constituent countries, from mid-1975; vi, 90p. of tables; 30cm; cover title has: 1975-2015 [0 11 690642 1] 350g; £2

Office of Population Censuses and Surveys *see also* General Register Office, Scotland

Office of Population Censuses and Surveys. Social Survey Division *see* Social Survey Division (Office of Population Censuses and Surveys)

Office of Public Information (United Nations)

The United Nations and outer space. New York: U.N.; iv, 42p.; 22cm [0 11 904972 4] 60g, £1.62, sold by HMSO [UN Sales no. E.77.I.9]

Office of the Health Service Commissioner *see also* Select Committee on the Parliamentary Commissioner for Administration (House of Commons)

Office of the Health Service Commissioner

Report, 5. Investigations completed. April to July, 1977; 138p.; 25cm (H.C. 505) [0 10 250577 2] 250g; £2.10

Office of the Parliamentary Commissioner *see also* Select Committee on the Parliamentary Commissioner for Administration (House of Commons)

Ordnance Survey

Annual report, 1976-77. – Department of the Environment (*Passive Author*) – Southampton: Ordnance Survey; 24p.; maps (1 fold), tables; 30cm [0 319 0000 36] 200g; £1.35, sold by HMSO

Organisation for Economic Co-operation and Development

Economic surveys

Canada. Paris: O.E.C.D.; 52p.; diagrs, tables(2 fold); 24cm [92 64 11660 5] 120g, £1.10, sold by HMSO [ISSN:0474-5183]

Germany. Paris: O.E.C.D.; 70p.; diagrs, tables(2 fold); 24cm [92 64 11635 4] 150g, £1.10, sold by HMSO [ISSN:0474-5183]

New Zealand. Paris: O.E.C.D.; 60p.; diagrs, tables (2 fold); 24cm: E; 24cm [92 64 11672 9] 150g, £1.10, sold by HMSO [ISSN:0474-5183]

The engineering industries in OECD member countries. basic statistics, 1972-1975. Deliveries of "100" selected products. Les industries mécaniques et électriques dans les pays membres de o'OCDE: statistqiues de base, 1972-1975. Livraisons de "100" produits individuels – Paris: O.E.C.D.; 94p.; mostly tables; 20 × 27cm – *in English & French* [92 64 01700 3] 250g, £2.50, sold by HMSO

The future of European passenger transport. final report on the OECD study on European inter-city passenger transport requirements – European Conference of Ministers of Transport & European Economic Community (*Collaborators*) – Paris: O.E.C.D.; 2v.

Figure 18: Sample page from and HMSO 'Monthly list'

(1918-), containing all important publications which appeared during the year. The annual catalogue is arranged in three sections: the first lists parliamentary papers numerically in their various series; the second is an alphabetical list of government departments and agencies with their publications, both parliamentary and non-parliamentary; the third section is a list of HMSO serials. Pagination is continuous within five-yearly periods, for which consolidated indexes are available.

In spite of the several indexes and catalogues listed above, government publications can still sometimes be elusive. A major problem is the difficulty of actually citing them correctly, and references are often found in contorted or abbreviated forms. One of the most common instances of this is the practice of referring to the reports of committees and commissions simply by the names of their chairmen; although there are several indexes for identifying these published by the Library Association Reference, Special and Information Section, it may be necessary to obtain professional assistance from a librarian or information officer to decipher the worst cases of this. Another problem is that of unpublished material, including items for internal departmental use, typescript reports, statistics, circular letters and memoranda, and a wide range of miscellaneous documents. Although unpublished, some of this is nevertheless available for use. A useful partial guide (and by definition it could hardly be comprehensive) is *Guide to government data: a survey of unpublished social science material in libraries of government departments in London*, A F Comfort and C Loveless (Macmillan, 1974), compiled for the British Library of Political and Economic Science.

US government publications

The Government Printing Office of the United States is possibly the world's most prolific publisher, and, as in the case of British official publications, there are several guides available to describe the different categories of material produced and the means of tracing individual items. Of these the most useful is *Government publications and their use* by L F Schmeckebier and R B Eastin (second edition, Brookings Institution, 1969). This is sensibly and logically arranged and includes separate sections on government periodicals and on maps. Depository libraries for US government publications are listed in an appendix. Older volumes which are still of some use are *United States government publications*, by A M Boyd and R E Rips (third edition, Wilson, 1949), which was in fact in print until quite recently, and *Manual of government publications; United States and foreign*, E S Brown (Appleton, 1950, and reprinted by Johnson Reprint in 1964). Two alternative more recent sources are *Subject guide to major United States government publications* by E Jackson (American Library Association, 1968, and the second volume of J B Mason's *Research resources: annotated guide to the social sciences* (American Bibliographical Center-Clio Press, 1971), entitled *Official publications: US government, United Nations, international organisations, and statistical sources.*

The basic current bibliography of American official documentation is the *Monthly catalog of United States government publications*, which has been published since 1895. The list is arranged under the names of issuing agencies and departments, and an index is provided of subjects, authors and titles. The indexes for each monthly issue cumulate annually in the December number. Congressional publications (including bills, reports and other material comparable in scope with the British parliamentary papers) are included and listed in a sequence under the general heading 'Congress'. An annual feature incorporated in the February issues is the *Directory of United States government periodicals and subscription publications.* A further, annotated list of serials is the three volume *Guide to US Government serials and periodicals*, J L Andriot, (Documents Index, 1969). For the period prior to the institution of the *Monthly catalog* there is a *Checklist of United States public documents 1789-1909*, published by the Government Printing Office in 1911. The British series of *Sectional lists* has an American counterpart in the official GPO *Price lists*, which identify documents in print; each list is based on a particular subject or group of subjects, and is revised at fairly frequent intervals.

Publications of international organisations

It was noted earlier that certain of the HMSO catalogues include the publications of international organisations, and there is an annual catalogue of *International organisations and overseas agencies publications* (HMSO, 1955-). A far wider range of associations is covered in the *Yearbook of international organisations* published by the Union of International Organisations; entries include not only details of activities and membership but in many cases brief reference to publications, mainly periodicals and series. The only comprehensive lists, however, are those produced by the organisations themselves.

United Nations documents index, published monthly since 1950, is the standard source for UN publications, although since 1962 the titles produced by specialised agencies of the parent body have not been included. There is an annual cumulation and subject index. For the period prior to 1950 there is *Ten years of United Nations publications, 1945 to 1955.* The documentation of the UN's predecessor, the League of Nations, is described in *Guide to League of Nations publications: a bibliographical survey of the work of the league* by H Aufricht (Columbia University Press, 1951).

The specialised agencies of the United Nations mostly issue their own bibliographies and catalogues of publications. The Food and Agriculture Organisation, for example, produces a *Catalogue of FAO publications* at two-yearly intervals, and there is the quarterly *FAO documentation: current index* (1967-). Unesco publishes a quarterly *List of Unesco documents and publications*. In 1973 a cumulated list was issued under the title *Bibliography of publications issued by Unesco or under its auspices, 1946-1971*, containing over five thousand items arranged by subject. A general description of the bibliographic structure of the United Nations and its associated organisations is *A guide to the use of United Nations documents* by B Brimmer and others (Oceana, 1962), although this is now somewhat dated. A more recent attempt to chart this complicated area is *Publications of the United Nations system: a reference guide* edited by H N M Winton (Bowker, 1972). The first part describes in turn the various agencies and their range of publications; the second is a list of selected documents, arranged by subject.

Other international organisations in general follow the same practice, and their own bibliographical lists provide a continuing review of publications of interest. These lists are traceable through general reference sources and bibliographies of bibliographies, and a few examples will suffice here. The European Communities issue an irregular *Publications of the European Communities: catalogue,* as does the Council of Europe: *Catalogue of the publications of the Council of Europe.* Details of EFTA publications appear in the *EFTA bulletin* (monthly, 1960-), and OECD issues a *Catalogue of publications* at intervals, which is kept up to date by supplements. These are typical of the bibliographies produced by international organisations, and these patterns of publishing are repeated by other similar bodies whose publications are also of interest to geographers. These official bibliographies can in some cases be supplemented by unofficial indexes and guides, which may in fact be easier to use; a good example is a recently published volume, *A guide to the official publications of the European Communities* by J Jeffries (Mansell, 1977).

Nine

THE HISTORY OF GEOGRAPHY AND GEOGRAPHICAL THOUGHT

THE TASK of concisely reviewing the development of geography from the earliest times to the present day, which is attempted in this chapter, is not an easy one. The period covered embraces many centuries, geographers of numerous schools and nationalities, and a very considerable volume of literature, comprising both primary and secondary sources. Of course any piece of geographical writing is an expression of contemporary geographical thought, and selection, in the case of primary materials, is very difficult. It should also be realised that many of the most significant developments in modern geography are methodological, and are discussed in chapter ten. A second problem in the selection of recent material is that it cannot easily be assessed without historical perspective: it is difficult, if not impossible, to identify works of lasting significance in these circumstances. For this reason there are very few recently-published books among the primary materials discussed in this chapter, which is concerned solely with history.

What follows, therefore, is a summary of the main trends and schools of thought, each section consisting firstly of an introduction to such secondary sources as are available, and secondly highlighting a few of those key works in each group which are established classics of the literature of geography. It is relevant at this point to emphasise that this is not only a very small selection, but it is a selection based on the importance of the literature. For example, while geographers in the western tradition must be aware of the significance of the Russian and East European contribution to current geographical philosophy, it is fair to comment that historically very few titles have had any real impact in the main stream of geographical thought, and fewer have been made readily accessible by translation. As sources on the history of Russian geography there is a paper by D J M Hooson, *The development of geography in pre-Soviet Russia* (in *Annals of the Association of American Geographers,* volume 58, 1968, 250-272), and secondly, a useful chapter in *All possible worlds:*

a history of geographical ideas, P E James (Odyssey Press, 1972), which briefly discusses developments from the last years of imperial Russia up to the 1960s. A Soviet view of Russian geography is presented in a collection entitled *Soviet geography: accomplishments and tasks* edited by C D Harris (American Geographical Society, 1962); this is a symposium of some fifty papers by Russian geographers on the history and character of Russian geography, embracing virtually every aspect of the subject. The direction of more recent work is indicated in *Soviet geographical studies* edited by G Abramov and Y Pivovarov, published under the auspices of the National Committee of Soviet Geographers in connection with the 23rd International Geographical Congress in Moscow in 1976.

General histories of geography

Some degree of historical background is incorporated in almost all of the general introductory textbooks on geography referred to in chapter five. Of particular value, however, especially for its account of the more recent evolution of geography, is *Geography in the twentieth century* edited by Griffith Taylor (third edition, Methuen, 1957). Part one provides a concise and readable account of the immediate past history of the subject. Similarly concise, but covering a broader period beginning with ancient and classical geography, are four further histories, all of which are arranged chronologically, which systematically describe the growth and development of the subject up to the present day. *Geography of geography: origins and development of the discipline* by R H Fuson (W C Brown Co, 1969) is an illustrated and popularly written history which is least satisfactory in respect of the modern period. *Histoire de la géographie*, by R Clozier (fourth edition, Presses Universitaires de France, 1967) is better balanced, but less detailed in terms of factual information. The third general history, which is the most detailed and still probably the best in the English language in spite of its age, is *The making of geography*, R E Dickinson and O J R Howarth (Oxford, 1933).

The most recent attempt at a fairly general (if not entirely comprehensive) history is *All possible worlds: a history of geographical ideas* by P E James (Odyssey Press, 1972); the greater part of this book is taken up with the main national schools of geography during the formative years of the first half of the twentieth century, but it also includes a concise and balanced account of both the classical antecedents and contemporary trends. *A question of place: the development of geographic thought*, by E Fischer, R D Campbell, and E S Miller (second edition,

R W Beatty, 1969) is a most useful collection of translated excerpts from many differing periods, countries, and languages. The book is organised chronologically, being 'primarily . . . a collection of ideas about geography in the order in which they were developed', and each quotation is prefaced by an introductory statement providing details about the author and some evaluation of his work. Most of the outstanding scholars who have made significant contributions to the development of geographical thought are represented. *The history of geography* is a selection of essays by J N L Baker (Blackwell, 1963), the majority of which are reprinted from the journals in which they were first published. A wide range of topics are covered in the various essays, all of course pertinent to the same central theme, but as is so often the case with such collections the book does not form a coherent whole, and cannot be read as such. The same reservation must be expressed also in relation to *Human nature in geography*, J K Wright (Harvard U P, 1966). This consists of a number of stimulating essays, most of which are reprinted journal articles, which refer 'to the impact upon geographical awareness (perception, cognition, knowledge, belief, study) of human emotions, motives and behaviour . . .' A partial history of a very different kind is *The history of the study of landforms* by R J Chorley, R P Beckinsale, and A J Dunn (Methuen). Two volumes have appeared so far (*Geomorphology before Davis*, 1964, and *The life and work of William Morris Davis*, 1973).

The more recent period, from the early nineteenth century, in which the foundations of modern geography were laid, is well covered in three authoritative and well-documented texts. *The makers of modern geography*, R E Dickinson (Routledge and Kegan Paul, 1969) discusses fairly briefly the pioneering work of Humboldt and Ritter, and goes on to outline in turn and in greater detail the development of the French and German schools which were the nurseries of modern geographical thought. *Regional concept: the Anglo-American leaders*, by the same author (and same publisher, 1976) is a complementary volume which discusses the contribution of British and American scholars. The book includes brief assessments of general trends but is largely devoted to a series of individual biographical studies which include numerous excerpts from both contemporary comment in obituary notices and from the original publications of the leading figures of each generation. *A hundred years of geography* by T W Freeman (Duckworth, 1961) approaches the same subject matter rather differently. Following several chronological introductory chapters is a series of chapters devoted to various physical, human, and regional branches of geography. A useful feature is an appendix of brief biographies of notable geographers of the period, each including references to obituaries; there are however some curious omissions from this list.

To these basic references it is worth adding *Modern geographers: an outline of progress in geography since 1800* by G R Crone, (third edition, Royal Geographical Society, 1970), which is a very concise but authoritative summary. *The geographer's craft*, by T W Freeman (Manchester U P, 1967) is a series of seven studies of individual geographers illustrative mainly of early twentieth century developments. A new and promising annual series published by Mansell is *Geographers: biobibliographical studies*, prepared under the auspices of the International Geographical Union Commission on the History of Geographical Thought. The first volume (1977), edited by T W Freeman, M Oughton and P Pinchemel, contains eighteen studies of well-known (and some lesser-known) figures, each with brief biographical details and a bibliography.

A fascinating study of the earlier history of geography is provided in *Traces on the Rhodian Shore: nature and culture in western thought from ancient times to the end of the eighteenth century* by C J Glacken (University of California Press, 1967). The main theme of this book is the relationship of human culture to the natural environment in terms of three ideas: the idea of a designed earth; the idea of environmental influence; and the idea of man as a geographic agent. There are extensive footnotes and references, and a bibliography. Also based on the exploration of a particular philosophical principle in the context of geographical thought is *Chorological differentiation as the fundamental principle of geography: an inquiry into the chorological conception of geography* by G de Jong (Groningen, J B Wolters, 1962). Starting from Ritter and Hettner, the theme is traced through the work of numerous later geographers, culminating with Hartshorne, with numerous quotations.

Hartshorne's own work, both as an historian of geography and as the author of probably the most influential philosophical contribution to geography of the twentieth century, demands inclusion here on both counts. *The nature of geography: a critical survey of current thought in the light of the past*, was originally published in the *Annals of the Association of American Geographers*, volume 29, numbers 3 and 4, 1939, and subsequently reprinted several times in monograph form. There is a bibliography of nearly five hundred items. The original conclusions presented were subsequently modified in the later work *Perspective on the nature of geography* (Rand McNally, 1959) and the two books should be considered together if Hartshorne's position is to be fully appreciated. The second volume to some extent improves the balance of the historical analysis in the first, which relied heavily on German sources.

Ancient and classical geography

There are a number of standard authorities on the early development of geography, several of which date from the later nineteenth century. The most comprehensive, and perhaps the most daunting, is *The dawn of modern geography* by C R Beazley (originally published in London, 1897-1906, and reprinted in the United States by Peter Smith, 1949). This is a scholarly and detailed study, with copious footnotes and references, in three volumes amounting to nearly two thousand pages. There are a number of reproductions of contemporary maps. Volume one covers the years up to about AD 900, and the subsequent volumes 900-1260, and 1260-1420, respectively. Its scope is thus somewhat greater than, for example, the second classic study of the origins of geography, E H Bunbury's *A history of ancient geography among the Greeks and Romans from the earliest ages till the fall of the Roman Empire* (originally published 1879, reprinted Dover Publications, 1959, with an introduction by W H Stahl). Unfortunately, the meticulous attention to detail which has established the authority of Bunbury (in two volumes and more than 1200 pages) does nothing to improve readability; while Bunbury remains the most important source for the period, a less elaborate study is likely to be more palatable as an introduction. There is, for example, *A history of ancient geography*, H F Tozer (published 1897, reprinted with notes by R Cary, 1935), or the rather more recent *History of ancient geography* by J O Thompson (Cambridge U P, 1948). For the Greek contribution the best source is *The classical tradition in geography*, C van Paasen (J B Wolters, 1957).

There is a discrepancy between the beginnings of geography and the beginnings of geographical literature. Bunbury, however, writes of Homer as the father of geography, and since both the *Iliad* and the *Odyssey* contain a certain amount of what might be termed geographical writing, and it is with literature that this book is concerned, this seems to be a reasonable starting point. Both are readily available in translated editions. Unfortunately much of the early work of geographical interest emanating from this period no longer exists in its original form. Among those whose writings have survived, however, are Ptolemy and Strabo, who both borrowed extensively from their predecessors and contemporaries and whose contribution to our knowledge of classical geography is therefore of considerable significance. Ptolemy's *Geographike huphegesis*

(Geographical guide) is translated into English in *Geography of Claudius Ptolemy*, edited by E L Stephenson (New York Public Library, 1932). Strabo's *Geography* is available in translation by H L Jones, *The geography of Strabo* (Heinemann, 1917) in eight volumes in the Loeb Classical Library.

Medieval and early modern geography

Geography in the Middle Ages by G H T Kimble (Methuen, 1938, reprinted Russell and Russell, 1968) is, although old, still the best general introduction to this period in the development of geography. Beginning with the passing of the classical era, the story is traced into the early years of the sixteenth century. There are a number of plates, illustrating early maps, and a short bibliography of both primary and secondary sources. Another elderly volume, also recently reprinted, is *The geographical lore of the time of the Crusades: a study in the history of medieval science and tradition in Western Europe* by J K Wright (American Geographical Society, 1925). The reprint is by Dover publications, 1965, and includes an introduction by C J Glacken. Again the starting point is classical, but the later medieval period is of course excluded. Both Kimble and Wright are scholarly and well-documented accounts which have stood the test of time and complement each other in providing a balanced study of the period.

Post-medieval geography, up to the first half of the nineteenth century when the foundations of modern geography were laid, is much less well-documented. For an introduction to this period one has to rely on more general histories. A notable exception to this generalisation is a recent study of one of the great names of the period, Immanuel Kant. *Kant's concept of geography and its relation to recent geographical thought*, by J A May (University of Toronto Press, 1970) is a detailed analysis of a pioneer of the modern philosophy of geography. Kant's major achievement, from the geographer's point of view, is his *Physische Geographie*, available most readily in any of the several collected editions of Kant's complete works. Before Kant, however, the gap between the classical period and the beginnings of modern geography is bridged by the work of several other pioneers, notably Peter Apian (1495-1552), Sebastian Munster (1489-1552), Bernhard Varenius (1622-1650), and Philip Cluverius (1580-1622). The major works of all of these writers are in Latin and are fairly rare and difficult to locate. There are no English translations in the case of Apian (*Cosmographicus liber*, 1524); Munster (*Cosmographia universalis*, 1544); and Cluverius (*Introductio in universam geographiam*, 1624). Varenius' *Geographia generalis* (1650) was translated by R Blome as *Cosmography and geography* (1693) and later, with additions and amendments by several editors including Sir Isaac Newton, as *A compleat system of general geography* (fourth edition, 1765). The first important geographical book to be written in English, Nathaniel Carpenter's *Geography delineated forth in two bookes* (1625) also belongs to this period.

German geography

It is with the German school of geographers that the modern history of the subject is usually said to begin. Useful summaries of the main personalities and ideas involved are available in Dickinson's *The makers of modern geography*, and in *Geography in the twentieth century*, edited by G Taylor, in a paper by S von Valkenburg which covers the more recent period. More detailed information is available in *Geographie: Europäische Entwicklung in Texten und Erläuterung* by H Beck (Verlag Karl Alber, 1973). Although ostensibly covering a rather wider field, about half of this book is devoted to the German contribution, with excellent full bibliographies giving ready access to the original sources.

Alexander von Humboldt (1769-1859) and Carl Ritter (1779-1859) were the dominant figures in what was perhaps the most significant and formative period in the evolution of modern geography. Humboldt travelled widely and wrote extensively, but the principal work on which his stature as a geographer depends is *Kosmos* (five volumes 1845-62), which laid the foundations for the development of physical geography. A recent and lively biography of Humboldt in English is *Humboldt and the cosmos* by D Botting (Michael Joseph, 1973). Ritter, by contrast, was primarily a human geographer. His major work, published 1817-8, was a monumental nineteen volume treatise, *Die Erdkunde im Verhältnis zur Natur und zur Geschichte des Menschen* ('Geography in relation to nature and to the history of mankind').

Among the second generation are Oscar Peschel (1826-75), author of *Neue Probleme der vergleichenden Erdkunde als Versuch einer Morphologie der Erdoberfläche* (1869) and a collection of papers entitled *Physische Erdkunde* published posthumously ten years later. Peschel is usually regarded as the founder of German physical geography, and was followed by Ferdinand von Richthofen (*Führer für Forschnungsreisende*, 1886, and a major regional study on China, 1877-1912), and Albrecht Penck (1856-1945), whose *Morphologie der Erdoberfläche* (1894) is still a standard text. Reference should also be made to the latter's son,

Walter Penck (1888-1923), whose *Die morphologische Analyse*, published posthumously in 1924 is available in translation as *Morphological analysis of landforms* (Macmillan, 1953, translated by H Czech and K C Boswell).

On the human side, Friedrich Ratzel (1844-1904), Richthofen's successor to the chair of geography at Leipzig, is the outstanding name. Ratzel was the author of a number of books and articles, of which the most influential are *Anthropogeographie* (two volumes 1882, 1891), *Völkerkunde* (three volumes, 1885-8), and *Politische Geographie* (1897). *Völkerkunde* is available in an English translation by A J Butler, with the title *History of mankind* (Macmillan, 1896-8). Fuller details of Ratzel's other writings are listed in *Friedrich Ratzel: a biographical memoir and bibliography* by H Wanklyn (Cambridge U P, 1961). Finally mention must be made of Alfred Hettner (1859-1941) who contributed two important theoretical studies on the philosophy and methodology of geography: *Das Wesen und die Methoden der Geographie* (1905), and *Die geographische Geschichte, ihr Wesen, und ihre Methoden* (1927). Many of his articles and papers may also be traced in the journal *Geographische Zeitschrift*, which he founded.

An influential German contribution to the modern development of settlement studies was *Die zentrallen Orte in Suddeutschland* by W Christaller, originally published in 1933 and available in an English translation as *Central places in southern Germany* (Prentice-Hall, 1966). The conceptual and methodological significance of this work was largely overlooked at the time, however, and German geographical thought was dominated for a number of years up to the Second World War by the concept of 'geopolitik' (basically the use of geographical argument in support of political policy), following the work of Karl Haushofer, whose *Geopolitik des Pazifischen Ozeans* was published in 1924. The influence of this theory was marked by the success of a specialised journal, *Zeitschrift für Geopolitik*, which published papers around this theme between its foundation in 1922 and 1955, and continued under various other titles right up until 1968. More recent work has laid great stress on regional study and has continued in the tradition of Ratzel to emphasise aspects of human geography, particularly social geography. Good examples are H Lautensach's study of the *Iberische Halbinsel* (1964) and *Sozialgeographie* edited by W Storkenbaum (Wissenschaftliche Buchgesellschaft, 1969), the latter being a collection of typical recent contributions in this field.

French geography

The French school is comparable with the German in stature and influence on the development of modern geography, and the same two general sources, Dickinson and Taylor, provide between them the most convenient summaries. A longer and more detailed review of the work of the French school is available in *Society and milieu in the French geographic tradition*, A Buttimer (Rand McNally, 1971). This volume, which is published in the Association of American Geographers monograph series, is an historical study of the French contribution to human geography. A concise and personal French view of the development of geographical thought in France is *Histoire de la pensée géographique en France* by A Meynier (Presses Universitaires de France, 1969). Reference may also be made to a collection of more than forty essays by French geographers illustrating and documenting the growth of this important school of geography. Entitled *La géographie française au milieu du XXe siècle* (Bailliere, 1956), it is edited by G Chabot, R Clozier, and J Beaujeu-Garnier.

Georges-Louis Leclerc, Comte de Buffon (1707-88), compiler of the massive *Histoire naturelle, générale, et particulière* (forty-four volumes), and Conrad Malte-Brun (1775-1826) are among the earlier writers, although the latter was not French by birth. Malte-Brun was the author of *Précis de la géographie universelle* (1810-29), which is avialable in translation as *Universal geography or a description of all parts of the world*...(1827-32). Frédéric Leplay (1806-82), although not a geographer as such, played an important part in the propagation in France of Ratzel's deterministic view of the relationship between man and nature.

The greatest name in French geography, however, is that of Paul Vidal de la Blache (1845-1918), whose work had a lasting influence both through his publications and through the whole generation of geographers who trained under him. Perhaps the best appreciation of his work can be gained from *Principles of human geography* (Constable, 1926 and later reprints) which is a translation of some of his collected writings, edited by E de Martonne. Among other works should be noted *Tableau de la géographie de la France* (1903), which typifies the regional approach which is characteristic of the French school, and his scheme for *Géographie universelle* (Colin, 1927-55), continued after his death by L Gallois. This was an ambitious enterprise to produce a definitive regional geography of the world, country by country. This also, in view of other similar grandiose projects by French geographers (for example, Malte-Brun and E Reclus) may be said to represent something of a French tradition. Elisee Reclus' *Nouvelle géographie universelle* (1876-94) has been translated as *The earth and its inhabitants* (1882-95), edited by E G Ravenstein.

Among Vidal de la Blache's pupils were a number who themselves contributed greatly to the high regard in which the French school is held. Perhaps the best known name is Jean Brunhes (1869-1930), largely for his *Géographie humaine: essai de classification positive*, which first appeared in 1910, and subsequently went through several editions and abridgements, and has been translated twice into English as *Human geography*. Maximilien Sorre (1880-1962) was the author of an important text in the same field, entitled *Fondements de la géographie humaines* (1952). Some years previously a classic work in physical geography was published by Emmanuel de Martonne (1873-1955), yet another of the master's famous students. This book, *Traité de géographie physique* (1909) also went into many editions, and in 1927 an English translation, *A shorter physical geography* by E D Laborde was published.

There are so many well-known geographers among Vidal de la Blache's pupils that selection for such a brief review as this is almost impossible. It might be appropriate, however, to mention, in addition to the above, Albert Demangeon (1872-1940) for his *Problèmes de géographie humaines* (1942 and later editions) and his study of the British Isles in the *Géographie universelle*, most familiar in the English translation by E D Laborde. It is worth noting in passing that the *Géographie universelle* contains examples of the writing of many more of the great French geographers whose works are not listed here individually.

American geography

Geography now and then: some notes on the history of academic geography in the United States by W Warntz (American Geographical Society, 1964) is a short history which, as the sub-title indicates, is exclusively directed towards the evolution of geography as an educational discipline in American schools and colleges. A broader view is taken in another volume from the AGS, entitled *Geography in the making: the American Geographical Society, 1851-1951*, compiled by J K Wright and published by the society in 1952. In portraying the history of the society this large volume with its lists of past officials and publications portrays also the history of American geography and discusses the ideas and achievements of many leading practitioners. Another volume with a dual role is *American geography: inventory and prospect*, edited by P E James and C F Jones and published for the Association of American Geographers by Syracuse U P, 1954. Again the two functions are summed up in the title: this is a series of review articles covering most of the main divisions of the subject, reviewing past work and 'the objectives and procedures of geographic research' as at the date of publication. Most of the papers have extensive lists of references.

The traditional father of American geography was Jedediah Morse (1761-1826), author of sermons and religious works as well as several regional studies of the United States, whose work is now generally regarded as of indifferent quality and little real significance. More important among a number of early regional studies is Thomas Jefferson's *Notes on Virginia* (1784). *The physical geography of the sea* (1855) by M F Maury (1806-73) is another early American contribution which has become something of a classic and has been reissued, newly edited and with an introduction by J B Leighly (Belknap Press, Harvard University, 1963). The greatest name in the development of physical geography in America is that of W M Davis (1850-1934). Davis wrote a large number of essays on both meteorological and geomorphological subjects, though it is for the latter that he is remembered: *Geographical essays* (1909) contains the more important of these and has been edited in a reprinted version by D W Johnson (Dover Publications).

In the field of human geography the influence of Ratzel was strong in the United States. A prime reason for this was the teaching of E C Semple (1863-1932) in her *Influences of geographic environment on the basis of Ratzel's system of anthropo-geography* (1911) and *American history and its geographic conditions* (1903). The latter was revised in collaboration with C F Jones in 1933, and both volumes are available in reprint (Russell and Russell, 1968). The environmentalist theories of Ratzel and Semple were developed further by E Huntington (1876-1947), who wrote prolifically and whose books included several best-sellers. The most influential of these (and most typical of Huntington's views) are *Mainsprings of civilization* (Wiley, 1945), and *Civilization and climate* (Yale U P, 1915). Both have been reprinted, the former by Mentor, the latter by Archon Books (Shoe String Press, 1971). Isaiah Bowman (1878-1950), head fo the American Geographical Society for twenty years, was another influential figure. His best known work *The new world: problems in political geography* (1919 and later editions) was a by-product of the Versailles peace conference and its timely publication ensured a significant impact throughout the world. An important landmark in the development of geographical thought was marked by the publication of *The morphology of landscape* (1925) by C O Sauer, one of the first outstanding statements on the nature of the discipline by an American geographer.

R E Dickinson, in *Regional concept* identifies four academic traditions of American geographers, and he pinpoints the publication of Hartshorne's *The nature of geography* in 1939 as the starting point of the third phase marking a decisive shift of emphasis from environmentalism to regionalism. The influence of Hartshorne's work is clearly evident among the contributors to *American geography: inventory and prospect* (referred to above) which was a kind of stocktaking by all the leading figures of this generation. In more recent years there has been a further shift of emphasis and in particular a concern for methodological issues, very much in line with developments in the United Kingdom (R J Chorley and P Haggett); good examples are the work of such scholars as W Bunge (*Theoretical geography*, second edition, 1966) and B L J Berry (*Spatial analysis*, 1968).

British geography

Although not conceived as a comprehensive and systematic history of British geography, *British pioneers in geography* by E W Gilbert (David and Charles, 1972) is a most useful collection of essays which provide a refreshing assertion of the role of British geographers in the development of the subject. As the preface points out, this contribution is given little recognition in most of the general histories. Many individual assessments may also be found in Dickinson's *Regional concept* (Routledge and Kegan Paul, 1976). The early period of British geography is covered in two volumes by E G R Taylor. The first of these, entitled *Tudor geography, 1485-1583* (Methuen, 1930) is concerned with the emergence of England as a seafaring nation and describes the voyages and discoveries of the time and their implications for the development of geographical knowledge. A large portion of the book consists of annotated lists of contemporary sources. The second volume, *Late Tudor and early Stuart geography, 1583-1650* (Methuen, 1934) is a sequel to the above and again includes an extensive bibliography of both published and unpublished primary sources.

Apart from the considerable contribution made to geographical knowledge by British travellers and explorers several names are featured among the pioneers in physical geography. Perhaps the most notable of these was Sir Charles Lyell (1797-1875), author of *The principles of geology*, originally published in 1830 and rapidly revised in several further editions. The academic study of geography in England owes a lot to Mary Somerville

(1780-1872), whose book on *Physical geography* was published in 1848 (and various editions to 1877, revised by H W Bates). The most renowned of British geographers, certainly in his capacity as a founder of academic geography, was H J Mackinder (1861-1947). In terms of published work his most lasting achievements were *Britain and the British seas* (1902), and *Democratic ideals and reality* (1919). A J Herbertson (1865-1915) was a contemporary of Mackinder and for a time worked as his assistant. He published titles in both physical and human geography, perhaps the most interesting of which is *Man and his work: an introduction to human geography* (1899), written in collaboration with his wife, F D Herbertson.

Important more as a compendium of factual information than as an original work of scholarship was *A handbook of commercial geography* by G G Chisholm (1889), which is incredibly, still in print, although now in its nineteenth edition (Longmans, 1975). Access to the work of H R Mill, another influential contemporary, is possible through his *Autobiography*, edited by L D Stamp (Longmans, 1951). French influence during this period is clearly evident in regional work such as that of M I Newbigin (*Mediterranean lands*, 1924, and *Southern Europe*, 1932), J F Unstead (*A systematic regional geography*, a series begun in 1935 several volumes of which are still in print, in revised editions), and P M Roxby, who was responsible for the three volume work on China in the Admiralty *Geographical handbooks series.* Among the leaders of the next generation were A G Ogilvie (1887-1954), author of *Europe and its borderlands* (1957) and editor and joint author of *Great Britain: essays in regional geography* (1928); and C B Fawcett (1883-1952), whose best known work is a small volume entitled *Provinces of England* (1919). Other branches of the subject which have recived special attention in the United Kingdom are historical geography (the best example being the work of H C Darby) and geomorphology: a good source of information on the latter field is *A bibliography of British geomorphology* edited by K M Clayton (Philip, 1964). In the last few years methodological issues have been dominant, and this work is discussed in the next chapter.

Discovery and exploration

The extension of geographical knowledge has traditionally followed two parallel courses, which can be categorised as theoretical and practical. Although most of the histories of the discipline described in this chapter are concerned principally with the evolution of the theoretical basis of geography, the practical aspect concerned with the discovery and description of countries, peoples, and places is an equally valid one. In many cases these trends are inextricably confused in the literature, but it is pertinent to identify some particular sources of information relating to discovery and exploration. Travellers' descriptive accounts are not only contributions to the sum of geographical knowledge, but by their methodology and outlook are sometimes also of significance to the conceptual framework.

A most valuable bibliographic guide to this literature is available in *A reference guide to the literature of travel, including voyages, geographical descriptions, adventures, shipwrecks and expeditions*, by E G Cox (University of Washington, 1935-38). Volume one covers the old world, and volume two the new; a third volume devoted to Britain was added in 1949, including 'tours, descriptions, towns, histories and antiquities, surveys, ancient and present state, gardening, etc'. Arranged chronologically within various subject and regional groups, there are numerous bibliographical and other annotations and indexes of names.

A second more recently published bibliography is the first volume of the National Maritime Museum Library catalogue, entitled *Voyages and travel* (HMSO, 1968). This comprises more than 1200 annotated entries, arranged primarily under regional headings, and represents all printed books on travel acquired by the library up to 1967. There are of course an immense number of individual reports and records of individual expeditions and voyages to all parts of the world, of greatly varying significance. Access is provided to many of the more important of these via the reprinted editions of the *Hakluyt Society* (which in various series have been appearing regularly since 1847 from Cambridge U P). In recent years it has been necessary to reprint (Kraus Reprint) yet again some of the earlier of the society's publications. This programme commemorates Richard Hakluyt's own famous compendium *The principal navigations, voiages, traffiques and discoveries of the English nation, made by sea or over-land, to the remote and farthest distant quarters of the earth, at any time within the compasse of these 1500 yeeres . . .* originally published in London, 1589. *The Hakluyt handbook*, edited by D B Quinn and published in two volumes by the society in 1974, contains essays on his life and work and a bibliographical guide to his writings.

Long established as the standard work on geographical exploration and discovery, *A history of geographical exploration and discovery*, J N L Baker (Harrap, 1937) forms a substantial single volume account, covering a period extending from classical times into the twentieth century. It remains the best introduction for the serious student. Also worth mentioning as a general history is a much more concise volume in the Hutchinson's University Library series: *Exploration and discovery*, by H J Wood (1951). This 'aims not at encyclopedic description but rather to interest the reader in major episodes in contrasted regions presenting contrasted problems'. For quick reference use the *Concise encyclopedia of explorations* by J Riverain (Collins, 1969) is a handy source of factual information. Translated from the French edition of 1966, there are short articles under the names of numerous explorers and places, with illustrations and maps.

Among several less academic but very readable popular histories of exploration are *Discovery and exploration: an atlas-history of man's journeys into the unknown* by F Debenham (second edition, Hamlyn, 1968), and *Explorers of the world*, W R Clark (Aldus, 1964). Both are lavishly illustrated in colour with maps and pictures, and within the limits set by their very evident attempt to appeal to the general reader are satisfactory as concise introductory texts. The most recent attempt at a dual purpose survey of this type is *The Mitchell Beazley world atlas of exploration* by E Newby (Mitchell Beazley, 1975). This is not an atlas as such, but a very fully illustrated general history designed to appeal to a mass audience. A further example of a similar type of book based on cartographic evidence is *Explorers' maps* by R A Skelton (Routledge and Kegan Paul, 1958). The maps are nicely reproduced and the accompanying text is at the same time scholarly and readable. The final example of books intended for the general reader is rather different, both in content and appearance: *The explorers: an anthology of discovery*, compiled and edited by G R Crone (Cassell, 1962) is a selection of extracts from accounts of travel in various parts of the globe, largely chosen for their readability, with brief introductory notes.

Beyond these general works are many more which are focused on particular explorers, particular places, or particular periods. Among the latter is a very good short history of the early period in *The ancient explorers* by M Cary and E H Warmington (revised edition, Penguin, 1963), and a more recent study entitled *Travel in the ancient world* by L Casson (Allen and Unwin, 1974), which spans the period from the early Egyptians to the sixth century AD. A much more detailed survey is Beazley's *Dawn of modern geography* (Smith, 1949). The next great era of discovery during the fifteenth, sixteenth, and seventeenth centuries is the subject of two important studies. Firstly, *Travel and discovery in the Renaissance, 1420-1620* by B Penrose (Harvard U P, 1952 and subsequent reprintings) is particularly useful for the bibliography which contains references to a great many primary sources including many of the Hakluyt Society volumes. Covering a rather broader period is *The age of reconnaissance* by J H Parry (Weidenfeld and Nicolson, 1963), which describes the course of exploration, trade, and colonisation, and analyses both causes and results. Parry has provided a supplement to this classic study in *The discovery of the sea* (Weidenfeld and Nicolson, 1975), which describes European maritime exploration up to and including the vital sixteenth century period. The story is continued nearer to the present day in *A history of geographical discovery in the seventeenth and eighteenth centuries,* by E Heawood (Cambridge U P, 1912, and reprinted by Octagon Books, 1965), and in *A history of African exploration* by D Mountfield (Hamlyn, 1976), a lavishly illustrated book concerned principally with the nineteenth century.

History of cartography

The history of geography, and of geographical travel, exploration, and discovery is intimately linked with the development of cartography. This is important both because the map is a distinctly 'geographical' medium and because much of the evidence for the study of early geography is of necessity cartographic. Skelton's *Explorers' maps* has already been cited in this connection. Although the maps themselves are beyond the scope of this study, some reference to the evolution of cartographic techniques is therefore pertinent at this point.

The main handbooks detailed below may be supplemented by use of *A guide to historical cartography: a selected, annotated list of references on the history of maps and map making* (second edition, Library of Congress Map Division, 1960). Compiled by W W Ristow and C E Le Gear, this lists nearly seventy references in several languages, with brief annotations. A second important reference source indispensible to the serious student of historical cartography is the periodical *Imago mundi: a review of early cartography* (1935-). This is an exceptionally well produced journal and the only one devoted exclusively to this field. The contents include bibliographies and reviews as well as articles of a high standard. Separately numbered supplements are issued from time to time.

Imago mundi was founded by L Bagrow, who is also the author of the best monograph review. Originally published in 1951 *Geschichte der Kartographie* has been translated into English and revised by R A Skelton as *History of cartography* (C A Watts, 1964). This is very well illustrated, and besides a select bibliography (amounting to nearly three hundred items, arranged by country) incorporates a useful list of cartographers to 1750, with brief biographical notes. A more extensive bibliography and generous bibliographical notes are printed in *The story of maps* by L A Brown (Little, Brown, 1949). This is very readable, and adequately illustrated, though less lavishly produced than Bagrow's book. A third general history planned on similar lines and again including a list of cartographers, is *Mappae mundi: die geistige Eroberung der Welt* by J G Leithäuser (Safari-Verlag, 1958). Perhaps the best of the comprehensive histories, however, is a new one, *Landmarks of mapmaking* by R V Tooley and C Bricker (Phaidon, 1977). After a brief initial discussion of general trends in the history of cartography, the continents are covered in detail, and the experience of Phaidon Press in the field of art publishing is used to good effect with some magnificent facsimile maps.

There are several shorter introductory texts, perhaps the best of which is *Maps and map makers*, R V Tooley (revised edition, Dawson Publishing/ Archon Books, 1978), which is unusually well-documented for such a concise survey. Most of the leading names are covered, from ancient times to the nineteenth century. *Maps and their makers: an introduction to the history of cartography* by G R Crone (second edition, Hutchinson University Library, 1962), is less well-illustrated, although sources of illustrations are listed in an appendix. Within its terms of reference, as an indication of 'the main stages of cartographic development', it provides a satisfactory introduction. So too does *Maps and man: an examination of cartography in relation to culture and civilization*, N J W Thrower (Prentice-Hall, 1972), which contains a broad overview of the history of cartography.

Among more specialised books on the history of cartography are *The mapmaker's art: essays on the history of maps* (Batchworth Press, 1953), which includes several interesting papers, particularly on English cartography. The history of the Ordnance Survey is described in *The early years of the Ordnance Survey* by Sir Charles Close (originally published 1926 and reprinted by David and Charles with an introduction by J B Harley, 1969) and for the more recent period in an official publication, *The history of the retriangulation of Great Britain, 1935-1962* (HMSO, 1967). Globes are covered in *Terrestrial and celestial globes: their history and construction* by E L Stevenson (Yale U P, 1921), which ranks as a classic on this aspect of the subject. A handy reference source for the study of both early maps and globes is *How to identify old maps and globes* by R Lister (Bell, 1965), which includes a list of cartographers, engravers, publishers and printers from about 1500 to 1850.

Ten

GEOGRAPHICAL TECHNIQUES AND METHODOLOGY

METHODOLOGICAL CONCEPTS are vitally important in geography. Since the subject matter of the discipline is so diffuse, it is in methodology that geography has traditionally found unity: the key factor is a synthetic approach via the idea of the 'region'. which finds physical expression in cartographic representation. There thus exists something of a confusion between theoretical thinking about the nature of the subject and concepts which are in fact methodological rather than philosophical. Naturally enough, this confusion is reflected in the literature. For example, a volume of essays already described (chapter five), entitled *Geography as human ecology*, edited by R Eyre and G R J Jones, is subtitled 'Methodology by example'. By contrast, an interesting collection of papers by S W Wooldridge, with a title which implies a methodological theme (*The geographer as scientist*, published by Nelson, 1956, and reprinted by Greenwood Press, 1969), has the sub-title 'Essays on the scope and nature of geography'.

The use of methodological characteristics in defining the nature of geography is becoming increasingly unsatisfactory. The quantitative techniques which play such an important role in modern geography are not essential to the nature of the subject: at least not in the same sense as are the more traditional processes of observation, recording, and analysis which are represented by, for example, fieldwork, survey, and cartography. The 'quantitative revolution', as it is sometimes called, is an application to geography of methods often conceived in other contexts, rather than a natural growth from within geography. Nevertheless there are statements on the nature of goegraphy based largely on mathematical method. One such is W Bunge's *Theoretical geography* (second edition, Royal University of Lund, 1966), published in the series *Lund studies in geography*. The book begins with a discussion of 'the nature of theory in science and what form scientific theory assumes when applied to geography. In attempting to provide a methodological answer, the general nature of science is explored . . . Following this general methodological approach to theoretical geography a great mass of substantive illustration is introduced and this evidence constitutes the bulk of the book.'

The same characteristic–philosophy explained in methodological terms–is found in a useful collection of papers reprinted from several journals under the title, *The conceptual revolution in geography* edited by W K D Davies (University of London Press, 1972). This brings together conveniently a number of important contributions from the past decade, organised in four sections: Geography and the role of ideas; Geography and the methods of modern science; Geography and the systems approach; and Geography and behaviour. *Explanation in geography* by D W Harvey (Arnold, 1969) is yet another example of the same linking of philosophical concepts, in this case the logic of explanation, with methodological ideas. Consideration is given to both traditional geographical methodology and to the more recent quantitative techniques, and the book was described by one of its reviewers as essential reading for 'all who seek the leitmotif of modern geographical research and teaching'. An extremely useful general statement on methodology which has recently become available in English is *Methods and perspectives in geography* by J Beaujeu-Garnier (Longmans, 1976); this was originally published in French in 1971 and the present edition was translated by J Bray. The International Geographical Union has established a Commission on Geographical Data Sensing and Processing which is concerned with all aspects of methodology. This body organised a major symposium in 1972 entitled *Geographical data handling*, and the two large volumes of its published proceedings (edited by R F Tomlinson) constitute an important contribution on geographical methodology, and include an extensive bibliography. For the most part, however, the literature on the methodology of geography is more specific, the various techniques being treated separately, and consideration of these groupings constitutes the remainder of this chapter.

Fieldwork

Geography is frequently described as an essentially practical subject, with its emphasis on the recording and subsequent analysis of direct original observations made in this field, and there are a number of guides to the techniques and methods employed in this process. It is unfortunate perhaps that many of these guides are written from the viewpoint of school-level geography, and the majority concentrate on the role of fieldwork as a teaching technique in academic geography. The result is that the subject is characteristically treated with extreme simplicity, although in many cases the material is equally applicable in geographical research at a more advanced level. One of the few exceptions to this trend is *Field study in American geography: the development of theory and method exemplified by selections* by R S Platt (University of Chicago, Department of Geography, 1959). As the title suggests, this is a selection of excerpts from field studies undertaken in America, with a commentary to explain their 'special significance as mile-posts in the development of field geography'.

Among the many books with an educational bias are some which emphasise techniques and some which consist largely of case studies, usually in the form of exercises. With the latter we are not concerned here, and in the case of the former only a small selection of the more useful are mentioned. A book which combines both features is *Geography in the field*, edited by K S Wheeler (Blond Educational, 1970). It begins by introducing fieldwork as a teaching technique in geography and contains a variety of chapters on observational work in differing types of landscapes. Similarly conceived as a teacher's guide is *Field work in geography* by P A Jones (Longmans, 1968), which includes sections on work suitable at various levels as well as advice on planning and financing. A third example, again very similar both in conception and in content, is *The purpose and organization of field studies* edited by M S Dilke for the University of Manchester School of Education (Rivingtons, 1965). This is the first volume of a series entitled *Field studies for schools*, later volumes of which consist of case studies in various regions of the British Isles. *A geographical field study companion* by P A Sauvain (Hulton, 1964) is slightly less obviously directed at the educational function of fieldwork, and although very concise and elementary is in some ways the most satisfactory of this group as a manual of practice. Fieldwork in higher education is the subject of a small pamphlet *Field training in geography*, published by the Association of American Geographers Commission on College Geography, 1968.

Map reading

Map reading is an integral part of fieldwork, and an ability to interpret data presented in cartographic format is part of the necessary equipment of the geographer. There are several books on this technique, including some which although published some years ago are still of considerable value. One such is *The geographical interpretation of topographical maps* by A Garnett (Harrap, 1930). This is an essentially practical work, with discussion of general principles centred around specific examples from maps in a supplementary atlas which accompanies the main volume. An even older (though recently reprinted) American text entitled *Interpretation of topographic and geologic maps, with special reference to determination of structure* by C L Dake and J S Brown (McGraw-Hill, 1925), is also worth a mention. No examples are included, and references are to maps of the United States Geological Survey; in fact the whole emphasis of the book is towards geology rather than geography. On the interpretation of physical data, however, this is a worthwhile contribution.

In the introduction of the above title it is observed that the only body suitably equipped to produce an adequately illustrated text on map interpretation is a government agency. In Britain this has been done by the Ministry of Defence in the army *Map reading manual* (new edition, HMSO, 1973). This covers photo interpretation as well as map reading, and is a well established standard work that has been revised and updated several times. Its primary function as a service manual determines the style, which contrasts very sharply with that of more academic texts listed below, but it is well illustrated with numerous maps, photographs, and line drawings, and it is clear and precise. Not an official publication, but lavishly illustrated with a generous selection of official maps from the Ordnance Survey, is *Techniques in map analysis* by B D R Worthington and R Gant (Macmillan, 1975).

As in the case of fieldwork, there are many elementary books on map reading intended principally for use at school level including for example, *Map reading and interpretation* by P Speake and A H C Carter (second edition, Longmans, 1970); *Map and photo reading: a graded course* by T W Birch (metric edition, Arnold, 1976); and *Reading topographical maps*, A H Meux (fourth edition, Hodder, 1970). All of these are adequate as an introduction. Undoubtedly the best up to date treatment for most purposes however, is *Map interpretation* by G H Drury (fourth edition, Pitman, 1972). Focusing attention on the representation of features of both physical and human geography on the map, this provides a systematic review of interpretation in various landscapes, and there is a short bibliography. *Map and landscape* by D Sylvester (Philip, 1952), is comparable in scope and usefully relates map reading to fieldwork, particularly in part three, 'The three-dimensional study of landscape'. An interesting French text is *Documents et méthode pour le commentaire de cartes (géographie et géologie)* by M Archambault, R Lhénaff and J-R Vanney (Masson, 1968). In each of the two volumes, the text is backed up by an excellent set of illustrative maps and diagrams.

Surveying

Practical geography is concerned with the making of maps as well as with their use, and although the construction of topographic and other maps is a very specialised and skilled operation, some understanding of the techniques involved is an important part of the geographer's equipment. The literature on this subject is considerable; much of it, however, is highly technical and is aimed at professional civil engineers and land surveyors. It is therefore appropriate to start with some of the shorter introductory texts.

There are in fact a number of concise texts written for student use which have run into several editions and are of proven value: *Surveying and levelling for students* by B H Knight (fourth edition, revised by H J Brend, C R Brooks, 1968); *Elementary surveying* by J Malcolm (third edition, revised by D H Fryer, University Tutorial Press, 1966); *Elementary surveying*, A L Higgins (third edition, revised by L A Beaufoy, Longmans, 1970). None of these texts is written specifically for geographers, but provide most satisfactory introductions to the basic principles of surveying. The topics covered are similar in each case, and there is little to choose between them. A further example of more recent origin which does refer in its introduction to geographical applications is *Fundamentals of plane surveying* by C D Goode (Butterworths, 1971). A small text by F Debenham, *Map making: surveying for the amateur* (third edition, Blackie, 1955, and regularly reprinted) must be included here as almost the only book on this subject written by a geographer, and intended for the amateur rather than the intending professional surveyor. *Surveying for field scientists* by J C Pugh (Methuen, 1975) is also intended for the amateur; the intention is to enable a complete beginner to produce a reasonable map working from simple observation and measurement with the most basic instruments.

Moving to slightly more advanced and more detailed texts, there are again some well-tried standard works which cannot be overlooked. *Practical surveying* by G W Usill (sixteenth edition, revised by K M and M P N Hart, Technical Press, 1973) was originally published in 1888 and is approaching its centenary. Comparable in general level and approach but including rather more information on such specialised aspects as geodesy and hydrographic surveying is a volume entitled simply *Surveying* by A Bannister and S Raymond (third edition, Pitman, 1972). Both of the above were written by civil engineers, and this is true also of *Surveying instruments and methods for surveys of limited extent*, P Kissam (second edition, McGraw-Hill, 1956). Like the previous examples this is conceived as a textbook, covering 'the fundamentals of surveying and those basic surveying procedures that make up the great bulk of surveying practice'.

Moving now to advanced texts, there is the two volume standard, *A treatise on surveying* by R E Middleton and O Chadwick (sixth edition, revised under the editorship of W F Cassie, Spon, 1955). Volume one covers instruments and basic techniques, and volume two deals with more advanced matters. From the geographer's point of view a treatment on this scale contains a lot of superfluous detail, and Middleton and Chadwick may be regarded as a reference work only. This is perhaps also true of *Practical field surveying and computations* by A L Allan, J R Hollwey, and J H B Maynes (Heinemann, 1968). Both works are reasonably well organised and indexed for use in this way. *Geodesy* by G Bomford (third edition, Oxford, 1971) is a comprehensive but specialised text accepted as standard in this field, and includes a substantial bibliography.

Cartography

Unfortunately for the English-speaking geographer, some of the best books on cartography are published in German. There is, for example, *Gelände und Karte* by E Imhof (Rentsch, 1968), which is exceptionally well-illustrated, and *Die Kartenwissenschaft: Forschungen und Grundlagen zu einer Kartographie als Wissenschaft*, M Eckert (de Gruyter, 1921-25). The latter work, in two volumes, is probably the most complete survey of the subject and an essential source for the advanced student, although it is now somewhat dated.

English language texts can be grouped in several categories. *Mapping* by D Greenhood (University of Chicago Press, 1964) is a book which, according to its introduction, was 'written to be read rather than studied', and was published in a previous edition with the title *Down to earth: mapping for everybody*. Similarly intended for a broad, non-specialist audience is *Maps and air photographs*, G C Dickinson (second edition, Arnold, 1976). This covers a wide field, including material on history, nature, uses and interpretation of maps, and, as the title indicates, aerial photographs. Other fairly elementary works which are more specialist in their approach include *Cartographic methods*, by G R P Lawrence (Methuen, 1971), a volume in Methuen's very useful *Field of geography* series. This makes a good modern introduction, concise and well illustrated. Fuller treatment at a similar level is available in a comparable American text, *Principles of cartography* by E Raisz (McGraw-Hill, 1962). A short bibliography is included.

Another work by the last mentioned author, entitled *General cartography* (second edition, McGraw-Hill, 1948), can be regarded as a systematic textbook rather than an introduction. Although still quite serviceable in many respects (for example in its convenient and appropriately illustrated section on the history of cartography) it is becoming dated, and has been replaced as the best general text by *Elements of cartography* by A H Robinson and R D Sale (third edition, Wiley, 1969). This is a comprehensive and well-organised survey incorporating material on modern photographic and automated techniques. A British work which is comparable in scope and function is *Maps, topographical and statistical,* T W Birch (second edition, Oxford U P, 1964).

A second British example, also widely accepted as a standard work, is *Maps and diagrams: their compilation and construction*, by F J Monkhouse and H R Wilkinson (third edition, Methuen, 1971). This covers a rather narrower field: there is nothing on the background–projections, survey, cartographic history–since this is basically a practical manual, covering equipment, materials and techniques relevant to different kinds of cartographic situation. Other useful and important books on special aspects of cartography include *Statistical mapping and the presentation of statistics* by G C Dickinson (second edition, Arnold, 1973), and *The look of maps: an examination of cartographic design* by A H Robinson (University of Wisconsin Press, 1966). The former illustrates some of the techniques of cartographic representation of statistical data, emphasising the importance of the choice of format if statistics are to be both displayed to advantage and at the same time not given a misleading slant; the latter examines more theoretically the visual and psychological implications of the detail of cartographic symbolism and style. *Maps and statistics* by P Lewis (Methuen, 1977) is not so much a guide to the techniques of presenting statistical data on maps as an introduction to statistical techniques with an emphasis on the use of maps in statistical analysis.

Map projections constitute a complex branch of the subject which has received separate attention in, for example, *An introduction to the study of map projections*, J A Steers (fifteenth edition, University of London Press, 1971). Originally published in 1927, this is a logical and systematic survey providing a clear explanation of a difficult topic. A more modern approach is found in another text at a similar level, *Map projections: an introduction* by P Richardus and R K Adler (Elsevier, 1972). Less elaborate and rather older introductory accounts worth noting are *An introduction to the study of map projection*, J Mainwaring

(Macmillan, 1943, and reprinted), and *Map projections*, G P Kellaway (second edition, Methuen, 1949, and also reprinted). *Coordinate systems and map projections* by D H Maling (Philip, 1973) is a good explanation of the underlying principles of the various projections, and their purpose.

Among a number of journals devoted to cartography are two important annuals: *World cartography*, published by the United Nations, which contains news of work and progress in various countries around the world; and *International yearbook of cartography*, published by G Philip under the auspices of the International Cartographic Association, which is a medium for scholarly articles in several languages. More frequent (two issues per annum) are *The cartographic journal*, a fairly typical society publication by the British Cartographic Society, with news items as well as articles, and *Surveying and mapping*, a quarterly professional journal from the American Congress on Surveying and Mapping.

The best source of current bibliographical information on cartography, additional to the general indexes and abstracting journals referred to in chapter three which cover the field selectively, is *Bibliographia cartographica* (Verlag Dokumentation Saur, 1974-). Each annual volume contains about two thousand entries arranged by subject, indexing about two hundred and fifty cartographic journals from all over the world. For earlier retrospective searching there is *Bibliotheca cartographica*, a classified list published twice yearly for the Deutsche Gesellschaft für Kartographie from 1957. Figure nineteen shows the similar arrangement used, with headings in German, English, and French; however, the format of the later works is much improved. *Bibliotheca cartographica* itself supplements an earlier monograph bibliography, *Die Kartographie 1943-1954: eine bibliographische Ubersicht*, by H-P Kosak and K-H Meine (Astra Verlag, 1955). *Bibliographie cartographique internationale* (annual, 1949-) is an international index of maps rather than cartography.

A useful source of reference to factual information is *Geography and cartography: a reference handbook* by C B M Lock (third edition, Bingley; Hamden (Conn), Linnet, 1976), which incorporates a revision of *Modern maps and atlases* (Bingley, 1969). The new work consists of alphabetically arranged entries (mostly fairly short and specific) for topics, publications, institutions, and people, with an emphasis on sources of further information. In this revised format, however, it no longer provides the coherent account of the subject that was the great strength of the former edition, with its narrative style: since it in fact refers the reader to *Modern maps and atlases* for certain topics it clearly does not supersede this work, which remains a useful, though inevitably dated, general introduction. There is a good short dictionary of cartographic terms entitled *Glossary of technical terms in cartography* (Royal Society, 1966), and the

771. V i n o g r a d o v, B.V.: Razvitie aerometodov pri geobotaničeskich issledovanijach v SSSR. In: Doklady Komis. aeros-emki i fotogrammetrii. Geogr.Obšč.SSSR. Bd 4. Leningrad 1967. S.33-48. (Die Entwicklung der Luftbildforschung bei geobotanischen Untersuchungen in der UdSSR. Russ.)

772. W i t t e n b e r g e r, Georg: Wissenschaftler und Amateure Hand in Hand. Die floristische Kartierung Mitteleuropas. In: Spessart. Monatsschrift des Spessartbundes. Aschaffenburg. 1968,12. S.12.

Zoologie / Zoology / Zoologie

773. M i r o n o v, N.P.:Principy i sposoby kartografirovanija poselenij malogo suslika v prirodnych očagach čumy. In: Gryzuny i ich ektoparazity. Saratov: Saratovsk.Univ.1968. S.446-451. (Prinzipien und Mittel der Kartierung von Siedlungen der kleinen Zieselmaus in natürlichen Pestherdgebieten. Russ.)

774. R u d e n č i k, Ju.V., L.A. K o p c e v, A.F. A l e k s e e v, A.S. S a b i l a e v u. D. K a r a b a l a e v: Opyt kartografirovanija poselenij bol'šoj pesčanki v severo-zapadnoj časti Kyzylkumov. In: Gryzuny i ich ektoparzity. Saratov: Saratovsk.Univ.1968. S.140-149. (Bericht von der Kartierung der Siedlungen der großen Rennmaus (Gerbillus tamariscinus) im Nordwestteil der Kysylkum. Russ.)

775. S n i g i r e v s k a j a, E.M.: Metodika i principy krupnomasštabnogo kartografirovanija naselenija grysunov tajgi Amuro-Zeijskogo plato. In: Voprosy geografii. 69,1966. S.127-146. (Methodik und Prinzipien der großmaßstäbigen Kartierung von Nagetierpopulationen in der Taiga des Amur-Zeija-Plateaus. Russ.)

776. Š v e d o v, A.P.: Novyj sposob sostavlenija kart obogaščenija fauny v kompleksnych atlasach. In: Doklad Instituta Geogr.Sibiri i Dal'n. Vostoka. 10,1965. S.32-35. (Ein neuer Weg zum Entwurf von Karten der Faunenbereicherung in komplexen Atlanten. Russ.)

Demographie / Demography / Démographie

777. (B i e n z, G.:) Bemerkungen zu einer Karte der Bevölkerungsdichte von Basel. In: Regio Basiliensis. 8,1967,2. S.177-179, 2 Pil., Schriftt.

778. B u c h a n a n, A.: Some problems of mapping population in urban areas at a national scale. In: Pakistan Geographical Review. 21(1),1966. S.41-45, 1 Abb.

779. D a s g u p t a, Sivaprasad: Some methodological problems of density mapping. In: Geogr.Rev.India. 26,1965,1. S.35-39.

780. D u p l e x, J., R. C o r n u u. L. R a o i l: Note sur l'Atlas sociologique de la France rurale. In: Revue Franç. de Sociol. Paris. 6,1965. S.127-136.

781. H ä r ö, A.S.: Area cartogram of the SMSA population of the United States. In: Ann. of the Assoc. of Amer. Geographers. 58,1968, 3. S.452-460, Abb., Ktn.

782. K o r o v i c y n, V.P.: O kartografirovanii naselenija. In: Materialy mežved. naučn. konferencii po Probl.narodonaselenija Srednej Azii,1965. Taškent 1965. S.46-47. (Über die Kartierung der Bevölkerung. Russ.)

Figure 19a: Sample page from 'Bibliotheca cartographica'

*Takahashi, Tadashi: Cartography of the age of exploration. Influence of the Islamic tradition on the Oriental geography. In: 7. Internat. Kongreß für Kartographie. - Madrid 1974. 6 p. 0288

Wallis, Helen: Working group on the History of Cartography. In: Internat. Geogr. Union. Bull. 25, 1974, 2. p. 62-64. 0289

Woodward, David: The study of the history of cartography: a suggested framework. In: Amer. Cartographer. 1, 1974, 2. p. 101-114, 5 ill., 5 tab. 0290

*Zavatti, Silvio; Franco Zavatti: L'Arcano del Mare di Robert Dubley. In: Universo. 53, 1973, 4. p. 695-712, 22 ill. 0291

B REGIONALE DARSTELLUNGEN UDC 528, 9 (091) (4/9)

REGIONAL REPRESENTATIONS

OUVRAGES RÉGIONAUX

Afrika / Africa / Afrique

Minow, Helmut: Praxis Geometriae - 5000 Jahre Vermessungswesen [2.] Antike Feldmesskunst in Ägypten. In: Vermess.-Ing. 24, 1973, 3. p. 83-86, 4 ill. 0292

Amerika / America / Amérique

*Birmingham, Ala. Public Library. 17th & 18th century maps of America by French cartographers: an exhibit of French map makers of America - 17th & 18th centuries from the Rucker Agee collection. - Birmingham: Birmingham Public Library 1973. 12 p. 0293

Cappon, Lester J.: Cartography and history: the atlas of early American history as a case study. In: Spec. Libr. Assoc. Geogr. a. Map Div. Bull. 94, 1973. p. 9-19. 0294

*Cartas de Indias. Tomo 1-3. [Repr. 1877]. - Madrid 1974. (Biblioteca de Autores Espanoles. 264-266.) 0295

*Chambers, Henry Edward: West Florida and its relation to the historical cartography of the United States. [Repr.] - Baltimore: John Hopkins Pr. 1898; [New York: Johnson Repr. 1973.] 59 p., mp. 0296

Clarke, J.; P.K. Macleod: Concentration of Scots in rural Southern Ontario 1851-1901. In: Can. Cartographer. 11, 1974, 2. p. 107-113, 3 mp. 0297

Clarke, J.: Military and United Empire loyalists grants in the Western District of Upper Canada 1836. In: Can. Cartographer. 11, 1974, 2. p. 186-190, 2 mp., 1 tab. 0298

Figure 19b: Sample page from 'Bibliographia cartographica'

International Cartographic Association has produced a *Multilingual dictionary of technical terms in cartography*, edited by E Meynen (Franz Steiner, 1973); this gives definitions in five languages (German, English, Spanish, French, and Russian) and lists equivalent terms in nine others.

Photogrammetry, air photo interpretation, and remote sensing

Photogrammetry, like other less specialised techniques of surveying, is a complex technical field and one which is peripheral rather than vital to the geographer. A full treatment is given in the *Manual of photogrammetry* (third edition, 1966, in two volumes) of the American Society of Photogrammetry. Rather more digestible is the more concise and readable treatment in two textbook-style summaries: *Photogrammetry*, F H Moffitt (second edition, International Textbook Co, 1967), and *Photogrammetry: basic principles and general survey*, B Hallert (McGraw-Hill, 1960). *Elementary air survey* by W K Kilford (third edition, Pitman, 1973), and *Elementary photogrammetry* by D R Crone (Arnold, 1963) are perhaps even more suitable for most purposes, and certainly as an introduction. Both books are written primarily for civil engineers, a group for whom the topic is similarly peripheral, and both treat the main principles clearly, without becoming too involved in a mass of technical detail. Another elementary introduction, written with the needs of geographers in mind, is *Simple photogrammetry: plan-making from small-camera photographs taken in the air, on the ground, or underwater,* J C C Williams (Academic Press, 1969). The International Training Centre for Aerial Survey in Delft is a prolific publisher, mainly of small pamphlets on technical topics. These provide modern statements on important themes and new developments which are applicable at an advanced level.

The interpretation of the results of aerial and other photographs is a matter of more general concern to the geographer, in several branches of his subject. Once again, the definitive work is produced and published by the American Society of Photogrammetry, *Manual of photographic interpretation* (1960). Should this prove too technical, there is *Aerial photographic interpretation: principles and applications*, D R Lueder (McGraw-Hill, 1959). Although this is published as one of a civil engineering series, it is a most useful treatment for geographers, covering in detail various landscape patterns and features, and further applications in geology, agriculture, forestry, urban studies and other fields. The best account for most purposes, however, is a recent one specifically written for geographers; this is *Geographical applications of aerial photography* by C P Lo (David and Charles, 1976). Several shorter texts are available which rely heavily on the use of illustrations. One example, *The uses of air photography: man and nature in perspective*, edited by J K S St Joseph (second edition, John Baker, 1977), consists of a series of short chapters on the application of photographic techniques in different disciplines. The book includes a section on *Air photographs and the geographer* by J A Steers, but the whole book is in fact of relevance. Another good introductory text, written as a 'self-instruction manual', is *Interpretation of aerial photographs*, T E Avery (second edition, Burgess Publishing Company, 1968). *Aerial discovery manual* by C H Strandberg (Wiley, 1967) is similar in content and approach. Special sections are devoted to photogeology and photohydrology.

Remote sensing is a relatively new field but already the literature is considerable. As noted in a previous chapter, a flood of papers in this area in the early seventies was enough to make it necessary to establish an additional section of *Geo abstracts*. A selection of these is listed, with abstracts, in *Radar remote sensing for geosciences: an annotated and tutorial bibliography* by M L Bryan (Environmental Research Institute of Michigan, 1973). A comprehensive introduction to remote sensing is available in *Remote sensing: techniques for environmental analysis* edited by J E Estes and L W Senger (Wiley, 1974). This is a collection of twelve papers on various aspects of the field, each with details of further reading, and the volume also contains a general bibliography and a directory of American institutions currently engaged in relevant research. *Introduction to environmental remote sensing* by E C Barrett and L F Curtis (Chapman and Hall, 1976) is a rather easier volume. It is a basic introduction to general principles, a review of the main techniques, and contains also some reference to applications. Much detailed technical information can be obtained and the development of remote sensing can be traced in the published proceedings of a series of symposia held at the University of Michigan beginning in 1962, and entitled simply *Proceedings of the third (fourth, etc) symposium on the remote sensing of environment.* Each volume is a massive compilation, comprising numerous papers on techniques and applications.

Photogrammetry and remote sensing are both areas experiencing rapid technological development, and it is essential therefore to note some of the principal journal literature in which these developments are recorded. *Remote sensing of environment: an international journal* is a newcomer, published quarterly since 1969 by Elsevier. Well established as the leading photogrammetric journal is *Photogrammetria*, also published by

Elsevier, the official journal of the International Society for Photogrammetry. This has been appearing since 1938, and contains authoritative articles and book reviews; special issues are sometimes devoted to longer review articles, such as that on *Photo interpretation applied to geomorphology*, volume 27, number 1, 1971. *The photogrammetric record*, a journal published twice yearly by the Photogrammetric Society, provides a forum for news, reviews, correspondence, and articles; each issue contains a section *International bibliography of photogrammetry* which indexes relevant literature published in other sources. Other useful journals which regularly contain relevant material are *Photogrammetric engineering and remote sensing* (American Society of Photogrammetry, 1934-), and *Photo interpretation* (bi-monthly, 1962-), published in Paris but with some contributions in English.

Quantitative methods

The application of mathematical and statistical techniques to geographical problems is very much in line with similar trends in other disciplines, and owes much to imported ideas from these fields. Correspondingly, some of the literature is more appropriately described in the context of specific branches of goegraphy rather than here, where we are still concerned with general sources. A second result of this interdisciplinary methodological development is that several works by scholars who are not primarily geographers are nevertheless central to the literature. The subject area perhaps most affected is economics, where the use of statistical techniques for the analysis of economic activity patterns can be broadened and applied to many locational and regional problems which are traditionally geographical. Perhaps the best examples of this development are in a group of books by the economist W Isard, notably *Methods of regional analysis: an introduction to regional science* (MIT, 1960), and *General theory: social, political, economic and regional, with particular reference to decision-making analysis* (MIT, 1969). Although much of the impetus towards quantification has come from external sources, some early examples written by geographers still occupy a prominent place in the literature: one such pioneering effort which has recently been translated from the original Swedish is T Hägerstrand's *Innovation diffusion as a spatial process* (University of Chicago Press, 1967, originally published in 1953). This edition contains a postscript by A Pred which sets Hägerstrand's work in context and explains its significance. Other early contributions are listed in *A working bibliography of mathematical geography*, M Anderson (University of Michigan, Department of

Geography, 1963), which owing to the very rapid pace of new developments in quantitative geography now looks very dated. Rather more useful is *A bibliography of statistical applications in geography* by B Greer-Wootten, published in 1972 by the Commission on College Geography of the Association of American Geographers. Various statistical applications are listed in turn, each with an introduction followed by a select list of references (over eight hundred in all).

A considerable number of systematic, shorter, and less complex texts mainly by British geographers, are available as an introduction. *Locational analysis in human geography*, P Haggett (Arnold, 1965) is one of the earlier of these and is a good example of the way in which statistical methods have evolved from location studies in economics into the broader field of human geography. The field is further broadened in a second volume by Haggett, on this occasion jointly with R J Chorley, entitled *Network analysis in geography* (Arnold, 1969). In this book the authors consciously build on the earlier volume and extend its methodological concepts to physical as well as human geography. The book explores 'ways in which the analysis of a topologically distinct class of spatial structures–linear networks–might throw light on common geographic problems of morphometry, origin, growth, balance and design' in both fields. The same two authors acted as editors for a collection of papers published as *Models in geography* (Methuen, 1967). This is a comprehensive review, introducing the role of mathematical models, discussing applications in physical, socio-economic and mixed systems, and information models. Each paper contains a list of references (often quite long) and the volume as a whole is an important reference source.

A comparable American reference on quantitative methods has provided two volumes in the series *Northwestern University studies in geography* (numbers 13 and 14). Collectively entitled simply *Quantitative geography*, these two volumes, edited by W L Garrison and D F Marble, are divided between economic and cultural topics (part one), and physical and cartographic topics (part two). The selection of papers is, however, less systematic than *Models in geography*, and includes several case-studies. A second American compilation is *Spatial analysis: a reader in statistical geography* edited by B J L Berry and D F Marble (Prentice-Hall, 1968). This is an important collection of essays on statistical methodology, regionalisation, and spatial statistics, distributions and associations. Collections such as these are a valuable asset where current research is proceeding with such speed that much of the really important material is found only in scattered journal articles. A more recent collection of stimulating but difficult papers from the prolific University of Bristol Geography Department is *Elements of spatial structure: a quantitative approach* by A Cliff, P Haggett, J K Ord, K Bassett, and R B Davies (Cambridge U P 1975).

Quantitative geography: techniques and theories in geography by J P Cole and C A M King (Wiley, 1968) is one of the best general surveys of the field, if somewhat elaborate as an introduction. Simpler and more suitable for this purpose are *Statistical analysis in geography*, L J King (Prentice-Hall, 1969); *Statistical methods and the geographer*, S Gregory (third edition, Longmans, 1973); and *Quantitative techniques in geography*, P Hammond and P S McCullagh (Oxford U P, 1974). Rather more concise, and perhaps more satisfactory for the beginner or those with a weak mathematical background are *The analysis of geographical data* by W H Theakestone and C Harrison (Heinemann, 1970), and *An introduction to models in geography*, R Minshull (Longmans, 1975). The most recent example is *Statistics in geography: a practical approach* by D Ebdon (Blackwell, 1976), which demonstrates basic concepts and makes extensive use of worked examples and practical exercises. The general application in geography courses at undergraduate level of sophisticated statistical techniques, which until quite recently were relevant only for advanced research purposes, has led to the production of this group of basic textbooks. They share a common approach in assuming only a minimal knowledge of mathematics and in emphasising the practical aspects of the uses of the methods described rather than the underlying theory.

Proficiency in mathematics can be a problem in the introduction of modern techniques; this is recognised in a basic account of the main areas of difficulty by A G Wilson and M J Kirby, entitled *Mathematics for geographers and planners* (Oxford U P, 1975). Naturally enough, many of the sophisticated quantitative techniques common in geographical research involve the use of computers and there are two recent textbooks on this area. *Computing for geographers,* by J A Dawson and D J Unwin (David and Charles, 1976) and *Computers in geography: a practical approach* by P M Mather (Blackwell, 1976) are both simple guides to the basics of computing and attempt to teach the main elements of programming. At a more advanced level, there is an annual publication which indexes research projects 'to promote the sharing of information on the use of computers and the availability of computer programs in the environmental sciences'. Published by *Geo abstracts* from the University of East Anglia, this began in 1970 under the title *Computers in geography,* and is now called *Computers in the environmental sciences.* Coverage is international, although the majority of entries are from Britain and the United States.

Eleven

PHYSICAL GEOGRAPHY

PHYSICAL GEOGRAPHY is a collective term for a group of individual specialisations within the discipline, each of which is the subject of a distinct body of literature. For the most part these specialised studies have close affiliations not only with geography but with several of the related disciplines in the physical and life sciences, which also contribute to the relevant literature. It is necessary, therefore, to be very selective in the choice of appropriate geographical texts, but at the same time to include a number of non-geographical items. References to the latter have in particular been kept to a minimum, and for the most part consist of standard bibliographies and bibliographic series, and guides to the literature of appropriate subjects. The inclusion of these enables the researcher to pursue a search for further materials, within the literature of fields related to geography but beyond the scope of this book.

With these considerations in mind, each of the major divisions of physical geography is discussed separately below (following some description of general sources), each section falling broadly into two parts. Firstly, there are references to certain categories of reference material, including basic current and retrospective bibliographies, periodicals, and guides to the literature of related fields. Secondly, there is a brief review of the more important general texts, emphasising those written by or for geographers, although not necessarily excluding others, and a few examples of more specialised texts devoted to particular themes of study.

General physical geography

The field of physical geography is outlined in a number of introductory textbooks, as well as in the *Larousse encyclopedia of the earth* (second edition, Hamlyn, 1972). This provides a comprehensive account of physical processes and phenomena affecting the earth, including earth history and the evolution of life forms. Characteristically, Larousse encyclopedias are very well illustrated, and this is no exception. A very different kind of reference source is *Physics and chemistry of the earth* (Pergamon, 1955-), which has had several combinations of editors. Originally intended as an annual volume, this contains in each issue several substantial review articles on a different group of topics, all bordering on geophysics and geochemistry. Terminology is explained in a very useful little work by W G Moore, *A dictionary of geography: definitions and explanations of terms used in physical geography* (fourth edition, Penguin, 1967). This has become established as a very popular and reliable source since its original publication in 1949, and is widely available. A more recent attempt is *Dictionary of the earth sciences* edited by S E Stiegeler (Macmillan, 1976), which brings together definitions from a rather wider range of disciplines. Many more useful general and specialised reference materials are listed in *The use of earth science literature*, edited by D N Wood (Butterworths, 1973). In common with other titles in the same series (*Information sources for research and development*), this is a mixture of thematic and discipline based chapters by librarians and subject experts, and both format and quality are variable. Rather narrower in its subject scope, but broader than its title suggests is D C Ward and M W Wheeler, *Geologic reference sources* (second edition, Scarecrow Press, 1972). There are four sections–general; subject (subdivided by topic); regional; and geological maps–and good indexes. Items listed are mostly annotated, some at considerable length.

Among the more successful of the textbooks are several written by A N Strahler, including some written jointly with A H Strahler. These include the well-established *Physical geography* (fourth edition, Wiley, 1975), and *Introduction to physical geography* (third edition, Wiley, 1973). Both books are very similar in general appearance and level, and are well-illustrated; the former emphasises geology and the latter meteorology and climatic factors. Both are adequate introductions for the non-specialist. More recently the same authors have produced two more substantially similar works: *Elements of physical geography* (Wiley, 1976), and *Principles of earth science* (Harper and Row, 1976). None of these volumes contains much of the mathematical material one might expect in a modern text in physical geography. A further useful American introduction is *Earth science* by R J Ordway (second edition, Van Nostrand, 1972), which is organised in four sections covering geology, astronomy, meteorology, and oceanography. *Physical elements of geography* by G T Trewartha, A H Robinson, and E H Hammond (fifth edition, McGraw-Hill, 1967) is yet another example of a very similar type, but is particularly strong on climatology and in contrast to the previous work contains only a brief statement on the earth as an astronomical body. The same authors have written *Fundamentals of physical geography* (third edition, McGraw-Hill, 1977) which is designed for a one-term introductory course.

Written at a slightly lower level, and more suitable for the sixth form than for undergraduate work, is *Introduction to physical geography* by H M Kendall et al (second edition, Harcourt, Brace, Jovanovich, 1974) and another work with the same title by R E Gabler, S Brazier, and R Sager (Holt, Rinehart, Winston, 1975). The latter emphasises the relationships between the earth's systems: land, water, and weather. A trend is apparent in a number of these introductory general textbooks which promote physical geography by emphasising applied aspects and links with fields of human geography. Typical examples are *Geology, resources, and society: an introduction to earth science* by H W Menard (Freeman, 1974), and yet another contribution from the Strahlers entitled *Environmental geoscience: interaction between natural systems and man* (Wiley, 1973).

British texts covering the whole field of physical geography are relatively hard to find. Perhaps the best known example, already in its eighth edition, is *Principles of physical geography* by F J Monkhouse (University of London Press, 1975). This is a systematic and balanced account, with numerous local instances quoted in illustration of the text, but the approach now looks somewhat dated. *The surface of the earth* by M J Selby (Cassell, 1967-71) is a very nicely illustrated two volume work, reminiscent in its physical appearance of the American examples already described. The first volume is concerned with the earth's structure and landforms, while the second and larger volume is in three parts, on climate, soils and biogeography respectively. The scope of *Principles of physical geology* by A Holmes (second edition, Nelson, 1965) is narrower, excluding the kind of material covered in Selby's second volume; it is more, however, than just a geomorphological source, and provides a useful geological background to a considerable range of topics in physical geography.

A useful French text by P Birot, originally published in 1959, is now available in an English translation by M Ledesert as *General physical geography* (Harrap, 1966). This is an interesting volume in which the several elements which comprise physical geography are explained in the context of their interrelationships, and without the divisions and distinctions which pervade most works on this topic. *Anatomy of the earth* by A Cailleux (Weidenfeld and Nicolson, 1965), translated by J Moody Stuart for the World University Library series, is a second French example, rather more limited in scope. This is a very concise, illustrated, summary which is useful as an elementary introduction, but not really comparable with the more comprehensive handbooks referred to above.

The majority of these texts are aimed at undergraduates beginning courses in the earth sciences, but are not intended as definitive statements applicable at a more advanced stage. In this respect they may be regarded as source books, covering the field logically and systematically, and they show many similarities both in content and style. *Symphony of the earth* by J H F Umbgrove (Martinus Nijhoff, 1950) is totally different in approach; this is a collection of seven papers on various topics, each of which is adapted from a lecture or lectures, and stylistically is very informal and readable. The text is supported by illustrations and a short list of references accompanies each paper. Also unorthodox in its general approach and method but in every other way unlike this collection is *Physical geography: a systems approach* by R J Chorley and B J Kennedy (Prentice-Hall, 1971). The systems approach referred to in the sub-title is a conscious break with traditional methods, and 'represents an unreserved attempt to show how the phenomena of physical geography can be rationalised and perhaps made to assume new significance and coherence when treated in terms of systems theory, statistical analysis, cybernetics, and other modern interdisciplinary approaches to the features of the real world'.

Geomorphology–general sources

The best starting point for research in the literature of geomorphology is the general guide by D C Ward and M W Wheeler already cited. *A guide to information sources in mining, minerals, and geosciences*, edited by S R Kaplan (Interscience, 1965) is a further source, particularly valuable as a directory of organisations, although much of its content, as implied by the title is rather marginal for geographers. There is a bibliographical list of publications, including many journal titles, in addition to the directory section. Two older guides which are now rather out of date but which should perhaps be mentioned in passing are *Guide to geologic literature*, R M Pearl (McGraw-Hill, 1951), and *The literature of geology*, B Mason (the author, 1953). In addition, a directory of institutions pertinent to several branches of physical geography exists in *Arid-lands research institutions: world directory*, P Paylore, published by the University of Arizona Press, 1967; and for North America there is a *Directory of geoscience departments* (American Geological Institute, 1970), which lists institutions, their activities, and their personnel. For historical and biographical information there are two source books of selected readings by major authors: *Source book in geology 1900-1950*, edited by K F Mather (Harvard U P, 1967), and for the earlier period *A source book in geology 1400-1900*, edited by K F Mather and S L Mason (Harvard U P, 1970, originally published by McGraw-Hill in 1939).

Several bibliographies, both current and retrospective, deserve some mention. The British Geomorphological Research Group have produced an annual *Current research in geomorphology* (1966-) and an earlier volume which covers thc period 1945-1962, *A bibliography of British geomorphology*, edited by K M Clayton (Philip, 1964). The closing date for this list ties in with the foundation of *Geomorphological abstracts*. As noted in chapter three, this was the predecessor of *Geo abstracts*, in which subsequent publications can be traced. A further (historical) source for British work is *The history of British geology: a bibliographical study*, J Challinor (David and Charles, 1971), and as a current bibliography the relevant sections of *Geo abstracts* and the other general abstracting/indexing services in geography can be supplemented by use of similar series in geology. These have had a complex history, but the most important work currently produced is *Bibliography and index of geology*, published by the Geological Society of America as a monthly abstracting service with an annual index volume. This service was previously known as *Bibliography and index of geology exclusive of North America* (1933-1968), but the scope has been increased and the title abbreviated. In its earlier format is was complementary to the US Geological Survey's *Bibliography of North American geology* (annual, 1896-) and *Abstracts of North American geology* (1966-1971), which formed a similar two-part service.

Two further series which have also ceased publication but which provide abstracts for a period otherwise covered only by the above annual indexes are *Geoscience abstracts* (American Geological Institute, 1959-1966), and *Geological abstracts* (Geological Society of America, 1953-1958). Aspects of the field applicable also in human geography are indexed in the semi-annual *Annotated bibliography of economic geology* (Economiç Geology Publishing Co, 1929-), which includes both monograph publications and articles in applied geology. *Earth science reviews*, a quarterly published by Elsevier, is a review journal which summarises recent advances on important topics, and is abstracted in *Geo abstracts*. *Geotitles weekly* is a current awareness service published by Geo-systems, a division of Lea Associates Ltd. Each weekly issue contains a classified list of more than one thousand references to recent literature. Geo Abstracts Ltd have recently revived an abstracting journal formerly produced by the United States Geological Survey (until 1971). This is entitled *Geophysical abstracts*, and in its present form dates from 1977. There are six issues a year, and it includes a fair amount of important report literature as well as covering the main journals. The largest single source for retrospective listing is the *Catalog of the United States Geological Survey Library*, published in 1965 by G K Hall, in twenty-five volumes.

Terminology is defined in several dictionaries, among which are *A Dictionary of geology*, J Challinor (fourth edition, University of Wales Press, 1973) and *Dictionary of geological terms*, C M Rice (Edwards, 1963). In addition there is the American Geological Institute's *Glossary of geology* (1972); and French and German terminology, as well as a limited selection of terms from other languages are covered in *Vocabulaire Franco-Anglo-Allemand de géomorphologie*, by H Baulig (Belles-Lettres, 1966), which is a classified list giving equivalent meanings, supported by an alphabetical index.

The Encyclopedia of geomorphology, edited by R W Fairbridge (Reinhold, 1968) is published as the third volume in the 'Encyclopedia of earth science' series, other volumes of which receive attention elsewhere in this chapter. This is a major work, with an impressive list of contributors. Each article is signed, all but the briefest of entries feature short bibliographical notes, and the whole is fully illustrated with both photographs and line drawings. Volume eight in this same series of encyclopedias was published in 1975; this is *The encyclopedia of world regional geology: Part I, the western hemisphere*, and it includes both Antarctica and Australia. A further encyclopedia, similarly one of a series, is *Standard encyclopedia of the world's mountains*, edited by A Huxley (Weidenfeld and Nicolson, 1962). Obviously much more limited in scope, this work might be more accurately described as a gazetteer, at least from the geographer's point of view. It includes human and historical data as well as physical description and statistical information.

Two important journals, one French and one German, are basic to the literature of geomorphology. *Zeitschrift für Geomorphologie* (1925-), although published in Germany and usually known by its German title, contains articles also in French and English. Each quarterly issue typically contains several substantial articles, often illustrated with large folded maps, plates or diagrams, shorter news reports, and a number of reviews. Each article is preceded by summaries in all three of the journal's languages. The French periodical is *Revue de géomorphologie dynamique* (1950-), currently edited by A Cailleux and J Tricart. Articles normally incorporate short summaries in English, and as well as book reviews each issue contains an abstracts section listing monographs and important papers in other journals. A third highly relevant title is *Geografiska annaler*, which is described in chapter four. Another new title worth noting, although it has not yet become fully established, is *Earth surface processes: a journal of geomorphology* (Wiley, 1976-), which could become an important forum for English language contributions to the literature.

The evolution of geomorphology as a science is the subject of *The history of the study of landforms*, a massive undertaking–three volumes are planned–by R J Chorley, A J Dunn, and R P Beckinsale. Volume one, published by Methuen in 1964, is sub-titled *Geomorphology before Davis.* Each section concludes with a bibliography, listing both primary and secondary sources. Volume two, on *The life and work of William Morris Davis*, was published in 1973. *Progress in geomorphology: papers in honour of D L Linton*, edited by E H Brown and R S Waters and published by the Institute of British Geographers in 1974, is an excellent collection covering most fields of geomorphology; each section is introduced by a useful editorial comment which helps to unify the volume, and the editors have also written a review of developments in British geomorphology since 1914. For geology, there is *The founders of geology* by Sir Archibald Geikie (second edition, Macmillan, 1905, reprinted by Dover, 1962), which covers the period from classical times till the late nineteenth century.

There are plenty of basic textbook introductions to the field, one of the most recent being *Fundamentals of geomorphology*, R J Rice (Longmans, 1977). This views the earth's surface 'as the interface of two energy systems, one fuelled by the sun and the other by internal sources within the earth' and is suitable at first year undergraduate level. *The study of landforms: a textbook of geomorphology*, by R J Small (Cambridge U P, 1970) is a sound modern textbook which provides a balanced and well-illustrated introduction, with useful lists of references to more detailed readings. A more comprehensive account is available in H F Garner's *The origin of landscapes: a synthesis of geomorphology* (Oxford U P, 1974); although concisely written, this is a massive volume of encyclopedic scope. *Principles of geomorphology* by W D Thornbury (second edition, Wiley, 1969) is the standard American textbook, originally published in 1954. An interesting feature is the final chapter on applied geomorphology, and each section includes short but pertinent lists of references, although many of the citations are rather old. A more recent review of the literature is contained in a substantial bibliography, pages 414-487 of *Introduction to geomorphology* by A F Pitty (Methuen, 1971), which is intended as a guide to the various schools and theories rather than 'to promote certain themes nor systematically to oppose more traditional studies . . . The sole conscious deliberation has been to illustrate . . . the nature of the actual, observed facts of landform study in preference to summarising the stockpile of hypotheses which have yet to be tested against facts yet to be observed.' A more theoretical treatment, relying heavily on the use of mathematics, is *Theoretical geomorphology* by A E Scheidegger (second edition, Springer, 1970).

Authors responsible for more than one work relevant to this selection include L C King, B W Sparks, and G H Dury. *The morphology of the earth: a study and synthesis of world survey* by King (second edition, Oliver and Boyd, 1967) is a massive tome covering the evolution of the earth from the mesozoic era to the present day, listing regional examples in groups under characteristic landscape forms and periods. The second of King's books is *South African scenery: a textbook of geomorphology* (third edition, Oliver and Boyd, 1963); the sub-title is more indicative of the importance of the work than is the title, although of course it is useful in both contexts. *Geomorphology* by B W Sparks (second edition, Longmans, 1972) is a more orthodox introduction to advanced study, written as a university textbook, and the same author has also made a useful contribution under the title *Roçks and relief* (Longman, 1971). In exploring the nature of rock resistance and its effect on structures, this study draws extensively on geological data and helps the geographer bridge the gap between the two disciplines relatively painlessly. There is something of a shortage of lighter introductory texts in geomorphology, but one such is G H Dury's *The face of the earth* (Penguin, 1960), which is both readable and authoritative, and is still in print after nearly twenty years. *Essays in geomorphology*, edited by G H Dury (second edition, Heinemann, 1968) is an important collection of original papers, most of which have regional associations.

A second example of a shorter than average introduction is *Geomorphology* by F Machatschek (Oliver and Boyd, 1969), translated from the ninth edition of the original German text. Concise but at the same time a reliable guide to a wide spectrum of topics and problems, it includes a useful bibliography which is particularly strong in French and German references. *The surface features of the land: problems and methods of geomorphology* (Macmillan, 1972) is a translation of Hettner's *Die*

Oberflächenformen des Festlandes (second edition, 1928) by P Tilley, who has also contributed a substantial preface explaining the significance of the work, and additional notes and references. This is hardly a textbook, but has been translated as a stimulant to English speaking geographers 'to do what Hettner induced his German colleagues to do half a century ago: review the present state of landform studies within our discipline, and in particular assess the work of W M Davis as a contribution to landform geography'. Unfortunately not available in translation, but an important reference and one which cannot easily be disregarded, even if only by virtue of its size and scope, is *Traité de géomorphologie*, by J Tricart and A Cailleux (SEDES, 1963-). It is planned in two parts, *Géodynamique physique,* and *Géomorphologie structurale*, each consisting of several volumes. Tricart is a prolific author, and several of his other books have appeared in English translations: one such which should be mentioned at this point is *Structural geomorphology* (Longmans, 1974), which was the first volume of his *Précis de géomorphologie*, originally published in 1968.

The methodology of geomorphology receives a balanced treatment in *Techniques in geomorphology*, by C A M King (Arnold, 1966), although in the light of current fashion statistical analysis is perhaps dealt with rather briefly. Each chapter includes a bibliography of further references. A fuller treatment of this aspect can be found in *Spatial analysis in geomorphology* (Methuen, 1972), an important collection of original papers edited for the British Geomorphological Research Group by R J Chorley; and in *Numerical analysis in geomorphology: an introduction* by J C Doornkamp and C A M King (Arnold, 1971). Statistical methodology, applied to geology rather than geomorphology, is the subject of *An introduction to statistical models in geology* by W C Krumbein and F A Graybill (McGraw-Hill, 1965), which nevertheless serves as a useful supplement to accounts provided by more general works on statistical method in geography (chapter nine). In view of the relative shortage of texts which concentrate on methodology, it is appropriate to introduce *Principes et méthodes de la géomorphologie,* J Tricart (Masson, 1965), a general introduction with an emphasis on techniques. The same trend towards an 'applied' approach that was noted in the general physical texts is also evident in individual branches of the subject; a good example is *Geomorphology in environmental management: an introduction* by R U Cooke and J C Doornkamp (Oxford U P, 1974), which relates geomorphology to the social, economic, and technical contexts where it can make a significant contribution.

Geomorphology–special topics; oceanography; hydrology

Among more specialised geomorphology texts are many devoted to particular landscape features, to the particular geomorphological processes by which these are formed, to particular geological periods, and of course to regional studies. A selection of some of the more important are detailed below. In addition there are fields such as hydrology and oceanography which are related to geography in the same sense as is geology, but which have not always been counted individually among the sub-disciplines of the subject. They are treated here as specialities within the field of geomorphology–an approach which can be justified on the basis of interrelationships within the literature.

All landscapes consist essentially of combinations of slopes, which therefore assume an important and central role in geomorphology. *Slopes: form and process* is the title of a selection of essays published as the Institute of British Geographers special publication number 3, 1971. Compiled by D Brunsden for the British Geomorphological Research Group, this is an important collection of original research papers by specialists, largely consisting of case studies. A more systematic approach is available in *Hillslope form and process* by M A Carson and M J Kirby (Cambridge U P, 1972), a volume in the series *Cambridge geographical studies*; and in *Slopes* by A Young (Oliver and Boyd, 1972), in the series *Geomorphology texts*. Many individual groups of landforms and landscape features have been documented at length. On coastal landforms the classic reference is *Shore processes and shoreline development*, D W Johnson (Hafner, 1919), which was reprinted in 1965. The text is well-illustrated and each chapter concludes with a summary and a short bibliography. Useful introductions to the field are *The physical geography of beaches and coastlines*, R K Gresswell (Hulton Educational, 1957), which is very concise, and *Applied coastal geomorphology* edited by J A Steers (Macmillan, 1971), a collection of reprinted essays. Steers has also produced an important regional study of coastal formations in *The coastline of England and Wales* (second edition, Cambridge U P, 1964), and its companion volume *The coastline of Scotland* (Cambridge U P, 1973). *Geographical variation in coastal development* by J L Davies (Oliver and Boyd, 1972) is a concise modern textbook which aims at a stocktaking of current thought. Among several more substantial works there is *Processes of coastal development*, V P Zenkovich, which introduces many Russian sources and examples, and is available in an English translation by D G Fry (Interscience, 1967). *Beaches and coasts* by C A M King (second edition, Arnold, 1972) is a similarly comprehensive and systematic survey by a British geographer.

There are useful lists of references at the end of each chapter. In addition to the citations made in the above two texts there are several bibliographies of beach and coastal features. *Annotated bibliography of quaternary shorelines* by H G Richards and R W Fairbridge (Academy of Natural Sciences of Philadelphia, 1965), was prepared for the seventh International Congress of the International Association for Quaternary Research, and covers material published 1945-1964. R Dolan and J McCoy have compiled a *Selected bibliography on beach features and related nearshore processes* (Louisiana State U P, 1965). This is arranged alphabetically under broad subject headings, and is not annotated. An elementary introduction to the same field is contained in *Coasts* by E C F Bird (MIT, 1969), which is a volume in the series *An introduction to systematic geomorphology.*

Uniform with this volume are *Karst*, J N Jennings (MIT, 1969), and *Volcanoes*, C Ollier (MIT, 1969), which again serve well as introductions. Both books include short bibliographies. Limestone topography is also the subject of two more notable recent contributions; *Karst*, edited by M Herak and V T Springfield (Elsevier, 1971) and *Karst landforms* by M M Sweeting (Macmillan, 1972). The latter makes an excellent introductory textbook and provides useful references for further study. Caves more generally are dealt with in *The science of speleology*, edited by T D Ford and C H D Cullingford (Academic Press, 1976), a varied collection of essays by twenty-two authors. Books on volcanoes tend to be rather spectacular and glossy in format, and many are more relevant to the geologist than the geographer, but there are thorough and scholarly works which are useful, such as *Volcanoes and their activity*, A Rittmann (second edition, Interscience, 1962), translated from the original German by E A Vincent; *Volcanoes* by G A MacDonald (Prentice-Hall, 1972); and *Volcanoes of the earth*, F M Bullard (University of Texas Press, 1976). The last mentioned is an excellent though non-technical introduction. The literature on coral formations includes several classics, notably *The coral reef problem*, W M Davis (American Geographical Society, 1928), and Darwin's *The structure and distribution of coral reefs,* originally published in 1889 and available in a reprint by the University of California Press, 1962.

On desert landforms a work of much more recent origin which has also achieved the status of a classic in its own field is *The physics of blown sand and desert dunes* by R A Bagnold (Methuen, 1954, but originally published in 1941). This is a neglected area, and there is not much literature readily available in English, but Bagnold's work should be supplemented by *Geomorphology in deserts* by R U Cooke and A Warren (Batsford, 1973) which is both readable and authoritative, and includes an extensive bibliography.

Glaciers and glaciation, like coastal landforms, have a considerable literature including several established standard texts from which selection is difficult. *Glacial and fluvioglacial landforms* by R J Price (Oliver and Boyd, 1973) and *Glaciers and landscape: a geomorphological approach* by D E Sugden and B S John (Arnold, 1976) are both sensible and well-written textbooks which serve as introductions to the field. More advanced, and incorporating a good selection of references to further sources, is *Periglacial processes and environments*, A L Washburn (Arnold, 1973). This is rather better illustrated than its main rival, *Glacial and periglacial geomorphology* by C Embleton and C A M King (Arnold, 1968); although still available in this well known 1968 edition, this has now been revised and reissued in two parts, entitled *Periglacial geomorphology* and *Glacial geomorphology* respectively (1975). *The periglacial environment* by H French (Longmans, 1976) is a more advanced treatment, concerned mainly with present day periglacial areas. Glaciation has its own journal literature: notable titles are *Journal of glaciology* (International Glaciological Society, three issues yearly, 1947-), and *Biuletyn peryglacjalny*, an irregularly published serial from Poland, in which most contributions are in French or English.

Several important texts are arranged around the concepts of various geomorphological processes. *The cycle of erosion in different climates* by P Birot and translated by C I Jackson and K M Clayton (Batsford, 1968) is an unusual study which re-examines and modifies the cyclic concepts of W M Davis. The basic erosional processes are described in general terms, and the influence of differing climatic factors is discussed in turn. *The mechanics of erosion*, M A Carson (Pion, 1971) is one of a series of *Monographs in spatial and environmental systems analysis* which emphasise modern techniques. The book consciously unifies and adapts to geomorphology concepts from disciplines beyond the experience of most geographers, in a more digestible form than some of the more advanced theoretical texts. *Fluvial processes in geomorphology* by L B Leopold, M G Wolman, and J P Miller (W H Freeman, 1964) 'deals primarily with landform development under processes associated with running water'. It is intended that the 'treatment of geomorphology in this book will provide a logical framework for the subject as a whole'. The same intention is implied in the title of a more recent book on the same theme *The work of the river: a critical study of the central aspects of geomorphology* by C H Crickmay (Macmillan, 1974). *Geomorphology and time* by J B Thornes and D Brunsden (Methuen, 1977) focuses attention on another critical factor in geomorphological explanation. This is an exploratory discussion rather than a definitive statement on the theme, and makes stimulating reading.

The process of weathering is covered in a concise volume entitled simply *Weathering*, C Ollier (revised impression, Longmans, 1975). This is a well-illustrated book with a useful bibliography of sources, which makes a very satisfactory introduction. *Properties of materials and geomorphological explanation* by W B Whalley (Oxford U P, 1976) is an unusual book, relating the raw materials of the earth's surface to the processes which form them into landscapes. *Geomorphology and climate*, edited by E Derbyshire (Wiley, 1976), is a collection of fifteen papers by an international group of contributors, aiming to present a statement of current thinking on the process-form relationship. A more systematic study, though now a little out-of-date, is *Introduction to climatic geomorphology* by J Tricart and A Cailleux (Longmans, 1972), and there is a useful but more limited text on a somewhat neglected aspect of this theme in *Tropical geomorphology: a study of weathering and landform developments in warm climates* by M F Thomas (Macmillan, 1974).

Predictably, the more important works relating to specific periods are geological rather than geographical, but several are useful for reference purposes. The Pleistocene is covered in *Glacial and Pleistocene geology*, R F Flint (Wiley, 1957) and the same author's *Glacial and Quaternary geology* (Wiley, 1971), which emphasise North America, and in *The Pleistocene period: its climate, chronology and faunal successions* by F E Zeuner (Hutchinson, 1959). This last contains a chapter on the Pleistocene chronology of the British Isles. *The Quaternary era, with special reference to its glaciation* by J K Charlesworth (Arnold, 1957) is a massive and detailed review in two volumes. The first of these discusses glaciology and glacial geology, the second the Quaternary. Volume two also contains a substantial bibliography, in addition to the very full lists of references which follow each chapter. *The ice age past and present* by B S John (Collins, 1977) is a more popular account of glaciated landscapes and environments, intended for the general reader. Also concerned with both past and present is *Arctic and alpine environments*, edited by J D Ives and R G Barry (Methuen, 1974), a massive and scholarly collection of eighteen essays dedicated to Carl Troll.

Oceanography is a difficult field from which to select items of importance to the geographer; although oceanography as a science is very much on the fringe of geography, the oceans occupy a large proportion of the surface of the earth, the totality of which is the study of geography. What follows is therefore a mixture of introductory texts, some of them designed for the use of geographers, and large reference compilations from which detailed information can be abstracted. Perhaps the best example of the latter group is *The sea: ideas and observations on progress in the study of the seas*, edited by M N Hill (Interscience, 1962-74). The five volumes cover Physical oceanography; Composition of sea water, Comparative and descriptive oceanography; The earth beneath the sea, history; New concepts of sea floor evolution; and Marine chemistry. It consists of a collection of papers by various authorities, each including a bibliography. The intention of this work was to provide an updated equivalent of *The oceans: their physics, chemistry and general biology* by H V Sverdrup, M W Johnson, and R H Fleming (Prentice-Hall, 1942). While this has been brought up to date by Hill's work it has not been fully superseded: in organisation it is undoubtedly more convenient, and it remains a standard reference source, and is still in print.

Reference works in the more usual sense of the term are even better organised for rapid verification of facts and definition of terminology, and there are oceanographic encyclopedias and dictionaries for this purpose. *The encyclopedia of oceanography*, edited by R W Fairbridge (Reinhold, 1966) is a useful and relatively up to date compilation in the 'Encyclopedia of earth sciences' series. Arrangement is alphabetical, backed up by a subject index, and each article is signed and includes a short bibliography. *Standard encyclopedia of the world's oceans and islands*, edited by A Huxley (Weidenfeld and Nicolson, 1963) reflects a very different concept. Islands, oceans, seas, and named coastal features such as headlands and bays are listed, with physical descriptions and some historical and other information. The technical terminology of oceanography is explained in *Glossary of oceanographic terms*, produced by the US Naval Oceanographic Office (second edition, Government Printing Office, 1966), which also includes a directory section.

The sea around us by R L Carson (revised edition, Oxford U P, 1961) is a concise and very readable introduction containing a short annotated bibliography. Also very readable, and assuming little detailed scientific background, is *Oceanography: the last frontier*, edited by R C Vetter (Basic Books, 1973), a collection of papers contributed by various eminent specialists and originally broadcast on American radio. *Descriptive physical oceanography: an introduction* by G L Pickard (second edition, Pergamon, 1975); *Principles of oceanography* by R A Davis (Addison-Wesley, 1972); and *The world ocean: an introduction to oceanography* by W A Anikouchine and R W Sternberg (Prentice-Hall, 1973) are all more academic texts, aimed at undergraduate level. *Oceans* by K K Turekian (second edition, Prentice-Hall, 1976), in the *Foundations of earth science* series, is a particularly valuable and up-to-date treatment suitable for the geographer not intending to specialise in this field. More detailed treatments written by and for geographers include *The physical geography of the oceans* by C H Cotter (Hollis and Carter, 1965), and *Oceanography for geographers* by C A M King. The latter has recently been revised and republished as two separate works: *Introduction to marine geology and geomorphology* and *Introduction to physical and biological oceanography*, both published by Arnold, 1975. Many aspects of oceanography of interest to geographers are discussed in *Submarine geology*, F P Shepard (third edition, Harper and Row, 1973).

Selection of appropriate sources in hydrology is subject to problems similar to those in respect of oceanography. Uniform with volumes already cited there is a *Standard encyclopedia of the world's rivers and lakes*, edited by R K Gresswell and A Huxley (Weidenfeld and Nicolson, 1965). Like the other Standard encyclopedias in this series, it contains a blend of physical and human information and in arrangement and subject matter resembles a rather detailed gazetteer. It is fully illustrated and contains an appendix of short entries in addition to the main sequence for which full information is provided. Also of some use but by no means a comprehensive dictionary is *Water and water use terminology* by J O Veatch and C R Humphrys (Thomas, 1966). The authors themselves state that they 'have been somewhat arbitrary and subjective in the matter of number and kinds of terms included . . .' The *International glossary of hydrology*, produced jointly by Unesco and the World Meteorological Organisation, will be a more satisfactory source. Preliminary drafts of the English language definitions of terms have been prepared, as have draft lists of equivalent terms in French, Russian, and Spanish. These drafts are for restricted distribution only at present. More widely available is *Elsevier's dictionary of hydrogeology in three languages: English-French-German*, compiled by H O Pfannkuch (American Elsevier, 1969). The main specialised bibliographic source is *Annotated bibliography on hydrology and sedimentation*, which has had a chequered history since its first publication in 1937 (under the title *Bibliography of hydrology*) by the American Geophysical Union. It now appears as part of the *Joint hydrology-sedimentation bulletin*, prepared for the US Interagency Committee on Water Resources.

The best textbook, written by a geographer for geographers, is *Principles of hydrology* by R C Ward (second edition, McGraw-Hill, 1975), intended as 'a straightforward, systematic analysis of the distribution and movement of water in the physical environment', and claims to be the 'first British textbook covering the general field of hydrology that is not aimed exclusively at the engineer'. Relevant sections of *Water, earth, and man* (chapter five) have been separately published and are also useful here. Other standard textbooks which can serve as reference volumes to the geographer include *Hydrology* by C O Wisler and E F Brater (second edition, Wiley, 1959). This work is an example of a text written for engineers rather than geographers; the same is true of a more recent and up to date alternative treatment in *Geohydrology* by R J M De Wiest (Wiley, 1965). *Handbook of applied hydrology: a compendium of water-resources technology*, V T Chow (McGraw-Hill, 1964) is a comprehensive source, well-suited to quick reference and including a detailed index. The scope is broader than the title suggests, and useful bibliographies accompany each section.

One of the more useful texts on special aspects of the subject of particular interest to geographers is *Streams: their dynamics and morphology*, M Morisawa (McGraw-Hill, 1968). This is a short and readable introduction to the dynamic principles guiding the activity of rivers covering not only hydrology and hydraulics but matters of more direct concern to the geomorphologist such as erosion, deposition, and the physical character of rivers and river systems. There is a useful introduction to the theme of ground water in *Ground water hydrology*, D K Todd (Wiley, 1959), in which further readings are detailed at the close of each chapter. There are several substantial bibliographies on ground water, among them two prepared for the US Geological Survey and published in the survey's 'Water supply papers' series: the first of these is a list of the survey's own contributions, *Bibliography and index of publications relating to ground water prepared by the Geological Survey and cooperating agencies* (by G A Waring, 1947); the second is also of limited scope, *Ground water in permafrost regions: an annotated bibliography* (by J R Williams, 1965). Flood problems are discussed in *Floods* by W G Hoyt and W B Langbein (Princeton U P, 1955), which deals with the causes and characteristics of flooding, and includes a bibliography. The examples, however, are mainly from the United States. There are several more modern treatments of aspects of flooding in the University of Chicago Department of Geography research papers series, including an interesting collection of essays edited by G F White under the title *Papers on flood problems*

(1961). The United States Office of Water Resources Research has published a bibliography on applied aspects of hydrology entitled *Bibliography on socio-economic aspects of water resources* by H R Hamiton (1966).

Meteorology and climatology

The most important source of bibliographical information in this field is *Meteorological and geoastrophysical abstracts*, formerly published under the title *Meteorological abstracts and bibliography* (1950-). This is a monthly, classified list with annual indexes, published by the American Meteorological Society. It is described by Walford as 'a model of its kind'. In the UK the Meteorological Office issues a monthly list of accessions, and older material may be traced through the Royal Meteorological Society's *Bibliography of meteorological literature*, which was published from 1922-1950 (two numbers per annum). The Meteorological Office has also produced a *Meteorological glossary* (fifth edition, HMSO, 1972), compiled by D H McIntosh. Other useful sources on terminology include *Glossary of meteorology* edited by R E Huschke (American Meteorological Society, 1959) and the more specialised *Illustrated glossary of ice and snow* by T Armstrong, B Roberts, and C Swithinbank (Scott Polar Research Institute, 1966). The latter of course has uses in other contexts, in addition to its meteorological value. Uniform with several other one-volume earth science encyclopedias by the same editor cited elsewhere is *Encyclopedia of atmospheric sciences and astrogeology*, R W Fairbridge (Reinhold, 1967). While not all of its content is relevant here, it maintains the same high standards seen in the others, and is a valuable reference source.

Important journals include *Monthly weather review* (US Weather Bureau, 1872-); *Journal of applied meteorology* (American Meteorological Society, bi-monthly, 1952-); *Bulletin of the American Meteorological Society* (monthly, 1920-); *International journal of biometeorology* (International Society of Bioclimatology and Biometeorology, quarterly, 1957-); *Meteorological magazine* (HMSO, monthly, 1866-); and *Weather* (Royal Meteorological Society, monthly, 1946-). All of these are abstracted in *Meteorological and geoastrophysical abstracts.*

Although there are available some handbooks designed to facilitate quick and easy reference to factual information in meteorology and climatology, these tend to be very dated. The two principal examples are *Handbook of meteorology*, edited by F A Berry, E Bollay, and N R Beers (McGraw-Hill, 1945), and *Compendium of meteorology*, edited by

T F Malone (American Meteorological Society, 1951). Both are still cited in relatively recent literature guides but although doubtless they still have their uses they should obviously be used cautiously. Alternatives exist in a number of up-to-date general textbooks although it must be admitted that few cover the field so comprehensively. Among the best of these are *Introduction to meteorology* by S Petterssen (third edition, McGraw-Hill, 1969), and *Meteorology* by W L Donn (third edition, McGraw-Hill, 1965). Neither of these presupposes a detailed knowledge of mathematics, and the latter places particular emphasis on marine aspects of the subject. In *Introduction to meteorology*, F W Cole (second edition, Wiley, 1975) the approach is more theoretical and although specifically written with arts students in mind, a greater reliance is placed on mathematical and physical explanation. However, there is still a great contrast between this and the treatment to be found in more advanced works such as *Introduction to theoretical meteorology*, S L Hess (Holt, 1959); the stated aim of this book is the painless introduction to those theoretical concepts from mathematics and physics which are an essential part of modern meteorology. A more recent attempt at the same purpose, giving the necessary theoretical background for advanced study, is *Atmospheres* by R M Googy and J C G Walker (Prentice-Hall, 1972).

The above are all meteorological rather than climatological, and the difference is a real one, although it is not particularly distinct in the literature. On climatology a useful and general standard reference is *Climatology* by A A Miller (ninth edition, Methuen, 1961), which introduces the major themes and provides a concise survey of climatic types by region. *Foundations of climatology: an introduction to physical, dynamic, synoptic, and geographical climatology* by E T Stringer, together with its companion volume *Techniques of climatology*, makes a serviceable and fairly comprehensive general introduction: both volumes were published by W H Freeman in 1972. Another basic text, which has proved very popular over a number of years, is *General climatology* by H J Critchfield (third edition, Prentice-Hall, 1974), which has three main divisions: physical, regional, and applied. This is perhaps the most readable and simplest of the group. A useful introduction which combines a modern approach with the avoidance of mathematical jargon is *Introduction to the atmosphere*, H Riehl (second edition, McGraw-Hill, 1972). This volume contains a number of illustrations reproduced from satellite observations. More concise than any of the above are two small volumes which also make good introductions: *Climate and weather*, H Flohn

(Weidenfeld and Nicolson, 1969), translated from the original German by B V de G Walden, and *World weather and climate* by D Riley and L Spolton (Cambridge U P, 1974). *Atmosphere, weather, and climate* by R G Barry and R J Chorley, which in previous editions was one of the best short summaries of modern climatology, has now grown into a rather more substantial text (third edition, Methuen, 1976). Barry has collaborated again, this time with A H Perry, to produce a further important volume on *Synoptic climatology: methods and applications* (Methuen, 1973). Both books are very comprehensive and scholarly studies, with excellent bibliographies. *World climatology: an environmental approach* by J G Lockwood (Arnold, 1974) is an advanced text assuming a reasonable grasp of both physics and mathematics.

Many of the above texts include regional climatic information, but some mention should be made of sources devoted exclusively or substantially to this purpose. *An introduction to climate* by G T Trewartha (fourth edition, McGraw-Hill, 1968) is in two parts, of which the second is concerned with the world pattern of climatic distribution. A second contribution from the same author, *The earth's problem climates* (University of Wisconsin Press, 1961), describes and explains a selection of interesting climatic features from various countries. A more systematic review is provided by *The climates of the continents* by W G Kendrew (fifth edition, Oxford U P, 1961); in *Climatology and the world's climates* by G R Rumney (Macmillan, 1968); and in K Boucher *Global climate* (English Universities Press,1975). All of these sources are relatively concise, single volume summaries, but the most important regional compilation is on a very different scale. *World survey of climatology* (Elsevier, 1969-) is a fifteen volume set under the general editorship of H E Landsberg, with specialist editors for individual volumes. The first four volumes consist of a general review of current thinking in the field, and the remainder of localised surveys each on a different geographical area. Not all of the volumes have appeared as yet, but eventually this should provide comprehensive and authoritative data on the climates of the world. *The English climate*, H H Lamb (second edition, English Universities Press, 1964) describes the weather of the British Isles.

Important but specialised topics include historical climatology and several areas of human application. The former is discussed in *Climates of the past: an introduction to paleoclimatology* by M Schwarzbach (Van Nostrand, 1963), edited and translated by R O Muir, which includes an extensive bibliography. *Climatic change: evidence, causes, and effects,* edited by H Shapley (Harvard U P, 1960) is a very readable collection of twenty two papers by various authorities presented at a conference on this theme. A second collection consists of papers by H H Lamb, under the title *The changing climate* (Methuen, 1966). Most of the items included have been published previously elsewhere. The same author has prepared *Climate past, present and future*, a multi-volume project of which one volume, entitled *Fundamentals and climate now* has so far been published (Methuen, 1972). The extent of interest in this specialised field is illustrated by the founding of a new periodical entitled *Climatic change* (quarterly, 1976-); this is described as 'an inter-disciplinary journal devoted to the description, causes, and implications of climatic change'.

Microclimatology is the subject of a classic study, *The climate near the ground* by R Geiger (Harvard U P, 1965). This edition is a translation of the fourth German edition of 1961: the work was first published in 1927. The agricultural implications of climate are discussed in such works as *Climate and agriculture: an ecologic survey* by J-H Chang (Aldine, 1968); *Methods in agricultural meteorology*, L P Smith (Elsevier, 1975); and in a symposium edited by J A Taylor under the title *Weather and agriculture* (Pergamon, 1967). An extensive list of further references to this field is a *Bibliography of agricultural meteorology*, compiled and edited by J Y Wang and G L Barger (University of Wisconsin Press, 1962), which contains nearly 11,000 entries. Weather forecasting and weather modification are subjects of increasing interest. *Human dimensions of weather modification*, edited by W R D Sewell is a series of papers prepared for a Symposium on the Economic and Social Aspects of Weather Modification held in 1965. The volume was published in the following year by the University of Chicago Department of Geography. A thorough account of one major concern of weather modification, of potential benefit in many parts of the world, is found in *Clouds, rain, and rainmaking* by B J Mason (second edition, Cambridge U P, 1975). A general discussion of forecasting is available in a concise introductory text, *The practice of weather forecasting* by P J Wickham (HMSO, 1970), published for the Meteorological Office. *Global weather prediction: the coming revolution*, edited by B Lusignon and J Kiely (Holt, Rinehart and Winston, 1970) consists of articles on the present capabilities and future prospects of remote sensing techniques in forecasting. E C Barrett, whose work on remote sensing is noted elsewhere in this book, has written a more systematic account of this development in *Climatology from satellites* (Methuen, 1973). Air pollution is a topical theme and is the subject of an extensive literature, much of it popular and much of it American. A suitably comprehensive review will be found in *Air pollution* edited by A C Stern (third edition, Academic Press, 1976-7), in three volumes, with essays contributed by various authors. A more comprehensive view of the general field of applied climatology is available in *Climate and the environment: the atmospheric impact on man* by J F Griffiths (Elek, 1976); and in *Principles of applied climatology* by K Smith (McGraw-Hill, 1975), which treats climate as a resource and outlines ways in which an understanding of climatic processes can improve its use.

Biogeography

Biogeography is concerned with the spatial distribution of plant and animal life, and with the relationships of living things to their environment. Essentially this involves three specialised fields within biology, namely phytogeography, zoogeography, and ecology. These are discussed separately below, but classifications and distinctions of this nature are artificial (if not superficial) and the biogeographer clearly requires access to some important general sources of information in biology. These are considered first of all.

In *Geo abstracts* series B covers the field selectively, and the other indexing journals also include relevant material. The majority of citations in these services, however, is from geographical sources, and a far broader spectrum of potentially useful materials are reviewed by *Biological abstracts* (1926-) which reviews about 8000 journals. This is a semi-monthly classified list, with annual indexes. Material of a peripheral nature omitted from the abstracts is indexed in the monthly *Bio-Research index* (1965-). *Biological and agricultural index* (Wilson, 1916-), formerly published (pre-1964) as *Agricultural index*, is a monthly bibliography rather restricted in scope by comparison with *Biological abstracts*, and is worth noting by geographers as much for its agricultural as its biological content. Useful quick reference sources covering the whole field of biology include *Biology data book*, edited by P L Altman and D S Dittmer (Federation of American Societies for Experimental Biology, second edition in three volumes, 1972-4), and *The encyclopedia of the biological sciences*, edited by P Gray (second edition, Reinhold, 1970). Articles in the latter are signed and include brief bibliographies. For terminology there is *A dictionary of biology* by M Abercrombie, C J Hickman, and M L Johnson (fifth edition, Penguin, 1966), which is adequate for the non-specialist user, and the much more detailed *A dictionary of biological terms* by J F and W D Henderson (eighth edition by J K Kenneth, Oliver and Boyd, 1963), which is the accepted standard work in the field. More specialised dictionaries of particular interest to the biogeographer are *Dictionary of ecology* by H Hanson (Peter Owen, 1962), and *Ecological glossary* by J R Carpenter (Hafner, 1962). Finally, for a comprehensive survey and discussion of biological information sources, embracing both biology in general and specific topics within it, there is *The use of biological literature*, edited by R T Bottle and H V Wyatt (second edition, Butterworths, 1971), and *Guide to the literature of the life sciences* by R C Smith and W M Reid (eighth edition, Burgess, 1972). The latter is best known as a zoological source only, but this latest edition has extended the scope. Review articles are published in *Biological reviews* (1923-), and *Quarterly review of biology* (1926-).

Biogeography: an ecological perspective by P Dansereau (Ronald Press, 1957) is a general text which 'extends across the fields of plant and animal ecology and geography, with many overlaps into genetics, human geography, anthropology, and the social sciences'. This is a well-illustrated volume with a useful bibliography, although the value of these references is restricted by the age of the book. More recent literature is cited in the bibliographies which accompany each chapter of *Biogeography: a study of plants in the ecosystem* by J Tivy (Oliver and Boyd, 1971). *Principles of biogeography: functional mechanisms of ecosystems*, D Watts (McGraw-Hill, 1971) is a modern and very readable introduction at undergraduate level. *The geography of life* by W T Neill (Columbia U P, 1969) is a profusely illustrated American text placing a greater emphasis on the description of distributions than ecological explanation. *Natural ecosystems* by W B Clapham (Macmillan, 1973) makes a good textbook suitable at first year undergraduate level, and *Biogeography*, H Robinson (MacDonald and Evans, 1972) is a simpler alternative. *Biogeography: an ecological and evolutionary approach* by C B Cox, I N Healey, and R D Moore was well-received when first published in 1973 and there is already a second edition available (Blackwell, 1976). It is written by biologists but expressly attempts to cater for a variety of approaches. The rapid increase in the popularity of biogeography in recent years is evidenced not only by the appearance of the above group of introductory texts, but also by the establishment of a major new journal, the *Journal of biogeography* (Blackwell, quarterly, 1974-), which contains research papers, shorter communications, and book reviews.

Perhaps the best all-round treatment of ecology is *Fundamentals of ecology*, E P Odum (third edition, Saunders, 1971), which is not only a comprehensive and systematic review of the field, but includes a forty-page bibliography of sources. This is unfortunately an alphabetical list only, where a classified arrangement would be much easier to use. *Introducing ecology* by G M Dunnett (Prentice-Hall, 1971); *Concepts of ecology* by E J Kormondy (second edition, Prentice-Hall, 1976); *Introduction to ecology* by P Colinvaux (Wiley, 1973); and *Ecology: the experimental analysis of distribution and abundance* by C J Krebs (Harper and Row, 1972) are acceptable alternative sources. Palaeoecology is the subject of *Principles of palaeoecology: an introduction to the study of how and where animals and plants lived in the past* by D V Ager (McGraw-Hill, 1963) and *Approaches to palaeoecology* by J Imbrie and N Newell (Wiley, 1964). *The ecology of invasions by animals and plants* by C Elton (Methuen, 1958) is a valuable ecological study of invasion patterns, both historical and recent. More advanced literature can be traced through *Ecological abstracts* (Geo Abstracts Ltd, bi-monthly, 1974-) which can be used to supplement the general biological abstracting and indexing services detailed above.

Plant and animal communities, both in terms of distributional and ecological studies, have received extensive separate treatment in the literature. *Zoogeography of the land and inland waters* by L F de Beaufort (Sidgwick and Jackson, 1951) is a concise survey based on a regional approach. An old but still useful and recently reprinted study on a much larger scale is *The geographical distribution of animals* by A R Wallace (Hafner, 1962, two volumes), originally published in 1876. *The distribution and abundance of animals*, H G Andrewartha and L C Birch (Chicago U P, 1954), *Ecological animal geography*, R Hesse, W C Allee, and K P Schmidt (second edition, Wiley, 1951), and *Animal ecology: aims and methods*, A MacFadyen (second edition, Pitman, 1963) are all basic sources: there are few works of more recent date which are comparable in scope. A useful introduction is found in *Animal ecology* by C Elton, originally published in 1927 but reprinted and available in paperback by Methuen, 1966. This edition includes a preface by the author pointing out the implications of recent advances. Not quite so old, but still in print and still useful, is *Animal geography: an introduction to vertibrate distribution* by W George (Heinemann, 1962). *Marine zoogeography* by J Briggs (McGraw-Hill, 1974) is one of the rare recent books in this field, but is obviously rather limited in scope.

The literature on the botanical aspects of biogeography is at least as extensive as that relating to animal life. *Geographical guide to the floras of the world* is a basic reference source which lists floras pertaining to all parts of the globe except Asia. It is an annotated bibliography of 'all the now useful floras and floristic works, including those published in periodical or serial literature'. Part one, by S F Blake and A C Atwood, published in 1942, covers Africa, Australasia, North and South America and the Pacific, Atlantic and Indian Ocean Islands; Western Europe is dealt with in part two, by S F Blake, 1961. Both are published by the US Department of Agriculture. There are very many more reference sources of value on plant life which could be listed, as is evidenced by the scope of the above bibliography, but among the most generally useful and readily available are the following: *The families of flowering plants* by J Hutchinson (third edition, Oxford U P, 1973); *A dictionary of the flowering plants and ferns* by J C Willis (eighth edition, Cambridge U P, 1973); and *Dictionary of economic plants*, J C T Uphof (second edition, Cramer, 1968).

Some of the best and most comprehensive texts on plant geography, unfortunately for the English-speaking geographer, are in German: *Die Vegetation der Erde in ökologischer Betrachtung* by H Walter (Fischer, 1962-8) is a most useful survey, covering the world in two large volumes. On a smaller scale, and in English, there is *The natural geography of plants* by H A Gleason and A Cronquist (Columbia U P, 1964), and *Foundations of plant geography* by S A Cain (Harper, 1944). *Plant geography upon a physiological basis*, A F W Schimper (Wheldon, 1960) is a second example of a German text conceived on a grand scale and with meticulous attention to detail, but unlike the work of Walter is available in English, translated by W R Fisher. Further basic works in English include *Introduction to plant geography and some related sciences*, N V Polunin (McGraw-Hill, 1960), and a more restricted text limited to selected plant groups, *The geography of flowering plants* by R D Good (fourth edition, Wiley, 1974). All of the above are lengthy volumes; shorter and more readable is *Plants and environment: a textbook of plant autecology* by R F Daubenmire (third edition, Wiley, 1974). The best introductory works for the non-specialist are *Plants, man, and the ecosystem* by W D Billings (second edition, Macmillan, 1972), and *Plant geography* by M C Kellman (Methuen, 1975) in the *Field of geography* series.

Soils

One of the most useful introductory works on the geography of plant life is *Vegetation and soils: a world picture* by S R Eyre (second edition, Arnold, 1968) which relates vegetation to soil type and soil formation, and which for this reason has been included here rather than in the preceding section. The treatment is regional, and an appendix lists the meanings of technical terms. The same linking of concepts is inevitably apparent in many of the other books discussed here: *Soils: an introduction to soils and plant environment* by R L Donahue (third edition, Prentice-Hall, 1971), and E J Russell, *Soil conditions and plant growth* (ninth edition, by E W Russell, Longmans, 1961) are but two examples in which this feature is apparent from the titles. The latter book is one of the more important standard references, and although there is no bibliography, there are numerous bibliographical citations in footnotes throughout the text.

Much of the most significant work on soils, from V V Dokuchaav (in the late nineteenth century) onwards, has come from Russian scientists. Several important and useful volumes are available in English translations. Three of these, all published by the Israel Program for Scientific Translations, Jerusalem, are V R Volobuev, *Ecology of soils* (1964), D G Vilenskii, *Soil science* (1963), and I P Gerasimov and M A Glazovskaya, *Fundamentals of soil science and soil geography* (1965). Volobuev includes an extensive bibliography, particularly strong on Russian sources. Both of the others follow a similar pattern, covering the physical, chemical and biological characteristics of soils, their formation and classification, and both include a regional description of soil types: in the case of the last mentioned title this extends to all parts of the world, while in Vilenskii this section is confined to the communist countries. In both cases these regional studies are very brief, with the exception of the USSR, which is discussed in considerable detail.

The nature and property of soils by H O Buckman and N C Brady (eighth edition, Macmillan, 1974) and *Fundamentals of soil science* by C E Millar, L M Turk, and H D Foth (fourth edition, Wiley, 1965) are both basic introductory texts with an emphasis on the agricultural significance of soils. *Soil* by G V Jacks (Nelson, 1954) is a concise and elementary outline which aims to 'place before the reader the latest findings on each aspect of the subject in as simple a manner as possible without diverging too far from the strict principles of scientific orthodoxy'. A second volume by Jacks, in this instance in collaboration with R O White, is *The rape of the earth: a world survey of soil erosion* (Faber, 1939 and later reprints). Although old, this is still a useful contribution to an important aspect which has applications in human as well as physical geography. A short regional analysis of soil types is *The geography of soil* by B T Bunting (new edition, Hutchinson, 1967) although curiously there are no maps to illustrate the distributions described. For more detailed consideration of the regional distribution of soils in Europe there is *The soils of Europe*, W L Kubiena (Murby, 1953). *Pedology*, *weathering, and geomorphological research* by P W Birkeland (Oxford U P, 1974) is a very useful study for the geographer, emphasising links between soil science and geomorphology and remarkable for its disregard of agricultural factors. *Soil geography* by J G Cruickshank (David and Charles, 1972) is an excellent and comprehensive undergraduate textbook with a particular emphasis on soil classification. *The soil: an introduction to soil study in Britain* by F M Courtney and S T Trudgill is more elementary and limited in scope, but makes a satisfactory introduction at first year level. The standard classification of soil types is that prepared and published by the Soil Conservation Service of the US Department of Agriculture, entitled *Soil classification: a comprehensive system* (seventh approximation, 1960). While this classification as published is only a tentative and provisional stage in the development of a truly definitive version, the latter has not yet appeared and this 'seventh approximation' must serve.

The current bibliography of soil science is best covered in primarily agricultural sources. The most important services are *Bibliography of agriculture* (United States Department of Agriculture, 1942-69), and *Bibliography of soil science, fertilizers, and general agronomy* (Commonwealth Bureau of Soils, 1931/4-). The former was a monthly indexing journal, arranged in subject sections, and having a worldwide coverage. An annual subject index was incorporated in each December issue. Soils, of course, represent only a small proportion of the entries. The printed version of the bibliography has now been replaced by a computer-based information service. The latter is published at three yearly intervals and this being so it is useful mainly for retrospective searching. It can also be used as a cumulative index to the abstracts published in *Soils and fertilizers* (1938-), one of a series of abstracting journals produced by the Commonwealth Agricultural Bureau. This includes review articles as well as the abstracts section, and is published six times per annum.

Twelve

HUMAN GEOGRAPHY

THE SCOPE of human geography is not easily defined, but broadly speaking it consists of those systematic fields of geography which are related to disciplines within the social sciences. The relationships involved are not however strictly comparable with those which link aspects of physical geography to the natural and physical sciences. Whereas physical geography appropriates to itself selected specific topics and themes of study from specialist outside disciplines, human geography takes potentially the whole range of social science data and analyses it geographically. It is difficult, therefore, to exclude material in the social sciences as irrelevant to the geographer. A second result of this complex relationship is that the classification of major fields within human geography is considerably more of a problem. While there are some clearly recognised divisions such as economic geography or political geography, there are in addition many minor special developments—medical geography, military geography, recreational geography—which do not always fit neatly within the main framework. In this chapter some of these themes are discussed together in the last section; others are treated as off-shoots of the major branches of human geography. Thirdly, there is confusion over terminology. Human geography, anthropogeography, social geography, and cultural geography are phrases which tend to be used very loosely, and although frequently regarded as synonyms have also been assigned distinct definitions in some contexts. The consequent interchange of this terminology in the literature will be apparent below.

Guides to the literature of the social sciences as a whole include several which discuss geographical sources, and these are described in chapter three. With reference to the more limited field of human geography, these same works are also useful for tracing information relating to other relevant individual subject fields in the social sciences. Further general social science reference sources which are pertinent to most of the divisions of human geography which are described separately below include *Encyclopedia of the social sciences* (Macmillan, 1930-5). Although this is now dated, it contains many still useful signed articles by leading authorities, incorporating many references to additional sources of information. The fifteen volumes span the entire field, and have not been superseded by the *International encyclopedia of the social sciences* (Macmillan, 1968). This is again a basic and authoritative source, published in seventeen volumes, with lengthy signed articles and short bibliographies. It is not a revision of the previous encyclopedia but a completely new work, and the two should be regarded as complementary. Brief definitions of terms are given in *Dictionary of social science* by J T Zadrozny (Public Affairs Press, 1959), and in greater detail in *A dictionary of the social sciences* edited by J Gould and W K Kolb (Free Press, 1964). Both of these are now rather dated, but there is a more recent work, *A dictionary of the social sciences* by H F Reading (Routledge and Kegan Paul, 1977); unfortunately this excludes economics, and gives only very brief, single phrase definitions.

An important retrospective bibliography covering the social sciences is the catalogue of the British Library of Political and Economic Science and holdings of certain other London libraries, entitled *A London bibliography of the social sciences.* Originally published in four volumes, 1931-2, this has been updated periodically by the publication of supplementary sections: each volume is arranged under subject headings, which are listed in the last volume of each set. The *International bibliography of the social sciences* is published regularly but is more useful for retrospective than current searching, owing to the long delay between the first appearance of each publication and its listing in the bibliography. Published annually in four series since 1952, this is a major bibliographic source which is genuinely international in scope. The four series cover sociology, politics, economics, and anthropology respectively, and each consists of a classified but unannotated list of books, articles, and other publications, with author and subject indexes. Originally produced by Unesco, the bibliography is now published by the Aldine Publishing Company, Chicago. *The ABS guide to recent publications in the social and behavioural sciences* is an annotated and select list of books and articles prepared by the journal *American behavioural scientist.* Originally published in 1965, it is updated by annual supplements.

General human geography

Since the several specialities within human geography have both multiplied greatly in number and become increasingly less part of a coherent whole, many of the best standard general introductions are now ageing, and can be counted with the literature of the history of geography. Their inclusion here is however essential, although it involves some repetition of parts of chapter nine. Mention must be made of Ratzel's *Anthropogeographie* and of the works of Ellsworth Huntington; in addition to the previously cited works he wrote a textbook of human geography, *Principles of human geography*, originally published in 1920, and more recently in a revised (sixth) edition by E B Shaw (Chapman and Hall, 1951). Further important American authors whose work is basic to this theme are E C Semple and C Sauer. Other classic texts include those by leading members of the French school, notably P Vidal de la Blache, J Brunhes, M Sorre, and A Demangeon. All of these scholars produced basic general works on human geography which are by no means of mere historical interest. (Fuller details of the relevant works are available in chapter nine.)

Continuing this French tradition are two more recent and up-to-date works, *Précis de géographie humaine* by M Derruau (Colin, 1963), and *Human geography* by A V Perpillou, translated by E D Laborde (Longmans, 1966). Both include bibliographies which are strong on French sources, and useful for this reason, in view of the importance of the contribution of the French school to this field of geography. More recently, P Claval has produced a further stimulating study based on the same continuing tradition and with an emphasis on the importance of historical factors, in *Eléments de géographie humaine* (Génin, 1974). Most of the other longer general texts are American in origin. *A geography of man* by P E James (second edition, Ginn, 1959) is a characteristic example which bears close resemblance to some of the general introductions to geography described in chapter five. The framework of the book is based on regions defined in physical terms; throughout, the physical and human features of each environment are balanced and related. By contrast *Introduction to human geography* by S N Dicken and F R Pitts (Blaisdell, 1963) is not regionally-orientated, but is arranged around a classification of major themes in human geography, each of which is systematically but briefly discussed in turn. *The human use of the earth* by P L Wagner (Free Press, 1960) adopts a similar approach, with a bias towards aspects of economic geography; 'the book shows man as the inhabitant and beneficiary of artificial environments, created by human effort and mediating between nature and the human individual'. *Man and the land: a cultural geography* by G F Carter (second edition, Holt Rinehart, and Winston, 1968), *Human geography: culture, society, and space* by H J de Blij (Wiley, 1977), and *Cultural geography* by J E Spencer and W L Thomas (Wiley, 1969) are more modern examples which provide good introductory statements covering the main themes. The last mentioned is interesting for its evolutionary approach and inclusion of a large amount of anthropological material. The same authors have produced a further more concise volume, *Introducing cultural geography* (Wiley, 1973). *Geography as social science* by G J Fielding (Harper and Row, 1974) provides a more theoretical approach, relating human geography to concepts from other branches of the social sciences; this may be supplemented for teaching purposes by a separate publication, *Programmed case studies in geography.* Perhaps the most balanced and the best all-round general summary, however, is *A geography of mankind* by J O Broek and J W Webb (second edition, McGraw-Hill, 1973).

All of the above are fairly large and substantial works which include a considerable amount of factual data, most of them being designed for beginning undergraduate use. Several more concise and readable introductions are also available, which exclude some of the detail but nevertheless provide a sound basic idea of primary concepts. *Human geography* by E Jones (Chatto and Windus, 1964) attempts to provide a logical framework for the subject, starting from population studies and proceeding through various forms of economic and social activity of increasing sophistication. More recently, Jones has collaborated with J Eyles in *An introduction to social geography* (Oxford U P, 1977) which is an excellent modern introductory text. *Situations in human geography: a practical approach* by J P Cole (Blackwell, 1976) is an unusual book on classroom techniques in human geography, based on learning by problem solving. *An introduction to human geography* by J H G Lebon (sixth edition, Heinemann, 1969) was originally published in 1952 but has undergone extensive revision in subsequent editions. The latest presents a substantially different view of human geography as 'the comparative study of major societies in the areas of their characterisation'. *An introduction to human geography* by D C Money (sixth edition, University Tutorial Press, 1975) is a simple and well-illustrated book, less philosophical than most of the former titles and aimed at a lower level. Also worth inclusion here, although it is in no sense a systematic human geography, is *Frontiers of geography* by O Hull (Macmillan, 1964). This sets out to introduce 'some of the topics that relate geography to everyday life', and discussion ranges widely over economic, social, political, and historical topics. *Human geography: a welfare approach* by D M Smith (Arnold, 1977) tackles the implications of applying geography to practical problems from a very different angle, focuses attention on one small field (welfare geography), defining it and relating it to geography as a whole.

The methodology of human geography is described in *Techniques of human geography* by P Toyne and P T Newby (Macmillan, 1971), which covers data collection, statistical methods, visual presentation of data, and location of studies. A further concise introduction is available in the excellent *Field of geography series* by R J Johnston entitled *Spatial structures: introducing the study of spatial systems in human geography* (Methuen, 1973). *Patterns in human geography: introduction to numerical methods*, D M Smith (David and Charles, 1975) is an attempt to provide a verbal rather than an algebraic explanation of quantitative techniques. *Human geography: evolution or revolution* by M Chisholm (Penguin, 1975) is not an introduction to techniques but a discussion and assessment of developments since the general adoption of quantitative methodology in the early and middle 1960s: it provides a good statement of what is now established theory.

Several collections of essays are useful and deserve a mention at this point. Perhaps the best known of these is *Readings in cultural geography* edited by P L Wagner and M W Mikesell (University of Chicago Press, 1962). Including numerous bibliographical references and representing a broad spectrum of geographical interests, this is a valuable volume containing papers both by geographers and by scholars in related fields. Human ecology is a phrase which has been used in contexts which suggest it as a definition for geography as a whole, and certainly it is appropriate as an indication of the scope of human geography. It is used in two collections of considerable interest to geographers: *Human ecology: collected readings*, edited by J B Bresler (Addison-Wesley, 1966), and *Studies in human ecology*, edited by G A Theodorson (Harper and Row, 1961). Both volumes consist of papers on a wide range of topics previously published elsewhere as journal articles, but their collection and juxtaposition make a valid contribution to the role of these volumes as sources in human geography. A recently published collection which surveys many aspects of a recurring theme in human geography is *Man's impact on environment*, edited by T R Detwyler (McGraw-Hill, 1971).

Man, space, and environment: concepts in contemporary human geography, edited by P Ward English and R C Mayfield (Oxford U P, 1972), is a useful source for sampling reprinted extracts illustrative of major themes and concepts in human geography, but taken as a whole fails to provide a coherent picture of the complete field. This objective is achieved more successfully in *Readings in social geography*, edited by E Jones (Oxford U P, 1975), which emphasises the problem of defining a satisfactory theoretical basis, but provides a classification of various approaches. *Studies in human geography*, edited by M Chisholm and B Rodgers and published by Heinemann for the Social Science Research Council (1973), is a volume of highly specific case studies reviewing recent work in various fields of human geography.

General human geography is related principally to fields such as sociology and anthropology, and bibliographic sources in these subjects provide the means of extending a search for relevant materials beyond the narrower confines of geography itself. In addition to the relevant series of the *International bibliography of the social sciences* and general indexing sources such as *Social sciences index, Humanities index* and *British humanities index*, several specialised tools within the above fields are also valuable. *Sociological abstracts* has been published since 1952 and currently appears as a classified list in six issues per year. More up to date information on new material is contained in the weekly *Current contents: social and behavioural sciences*, which reproduces the contents lists of current journals. *Current sociology* (three parts per annum, 1952-) provides reports and bibliographies on progress on selected topics in each issue. The Royal Anthropological Institute of Great Britain and Ireland issues a quarterly *Index to current periodicals received in the library* (1963-) which lists the contents of a large number of relevant journals, mainly under a regional classification. *Biennial review of anthropology* (1959-71) contained regular review articles of a high standard on selected topics which cover a wide range of anthropological (and geographical) interests: this has now been superseded by *Annual review of anthropology* (1972-). *Abstracts in anthropology* (Greenwood, quarterly, 1970-) groups abstracts of journal literature in four sections, on archaeology, ethnology, linguistics, and physical anthropology.

Population geography

A bibliographic guide to population geography by W Zelinsky, is a volume in the University of Chicago Department of Geography research papers series, published in 1962. It attempts to be a comprehensive listing to 1961, and contains more than 2500 unannotated citations, all but a few hundred of them regional. Although this list is now dated, it can be supplemented by sources such as *Population index*, published quarterly by the Population Association of America and the Office of Population

Research at Princeton University. The bulk of each issue consists of a classified sequence of short abstracts, with author and locational indexes. Also included are short articles, news items, and statistical tables. A further bibliographical source is the French journal *Population* (Institut National d'Études Démographiques, 1946-). Published bimonthly, every issue contains a large number of brief reviews of books, and abstracts of other material, in addition to several substantial articles on demographic topics. Other important general journals in the field include *Population studies* (Population Investigation Committee, London School of Economics, quarterly, 1947-), and *Demography* (Population Association of America, quarterly, 1965-).

As well as the *Bibliographic guide*, Zelinsky has written *A prologue to population geography* (Prentice-Hall, 1966), which is perhaps the best introductory text on the subject. In three parts, it defines the scope of population geography, describes the world distribution of population and discusses the concept of population regions. Other short general accounts include *Population geography*, M G A Wilson (Nelson, 1968) and *A geography of population: world patterns*, G T Trewartha (Wiley, 1969). *Spatial population analysis* by P H Rees and A G Wilson (Arnold, 1977) is a more modern and statistically complex study. Its 'main concern is to integrate the fields of spatial analysis and demography in the study of population and population change'. *Population geography* by J I Clarke (second edition, Pergamon, 1972) is a systematic rather than a regional text, but aims to suggest a framework on which to base regional studies of population. *Geography of population* by J Beaujeu-Garnier (Longmans, 1966, translated by S H Beaver) is a standard reference based on research carried out in France: the original work was published in two volumes, 1956-8. Among the other examples of valuable French references are several by P George, notably *Questions de géographie de la population* (Institut National d'Études Démographiques, Presses Universitaires de France, 1959); also worth noting is *Géographie de la population* (Presses Universitaires de France, 1965). This small volume in the *Que sais-je?* series is a very concise but authoritative summary.

All of the above are consciously geographical contributions, but as in most areas of geography many more 'external' sources are equally important. *Principles of demography*, D J Bogue (Wiley, 1969) is a thorough and well-illustrated general text, and in common with most of the relevant non-geographical works is organised thematically rather than regionally. This is true also of *Population*, W Petersen (third edition, Macmillan, 1975), although a considerable use is made here of regional examples. There are useful bibliographies at the end of each chapter. *Population problems* by W S Thompson and D T Lewis (fifth edition, McGraw-Hill, 1965) is a basic source of long standing, the first edition being published in 1930; again the treatment is thematic, with most examples taken from the United States. *A workbook in demography* by R Pressat (Methuen, 1974) is an English translation of a French university textbook, originally published in 1966.

There are a number of useful collected sets of papers on population problems, both geographical and non-geographical. An example of the former is *Population geography: a reader*, edited by G J Demko, H M Rose, and G A Schnell (McGraw-Hill, 1970). This consists of thirty-five reprinted essays grouped under several general themes, each of which is briefly introduced by the editors' commentary. Some specifically geographical sections are included in *The study of population: an inventory and appraisal* edited by P M Hauser and O D Duncan (University of Chicago Press, 1959), an important collection of review articles covering all aspects of the topic. O D Duncan is also joint editor (with J J Spengler) of *Demographic analysis: selected readings* (Free Press, 1956). This includes both a section devoted to regional studies and a substantial bibliography. *Population in perspective*, edited by L B Young (Oxford U P, 1968) is an anthology of mostly very short papers 'brought together to help the reader achieve perspective on the population problem both in historical depth and in breadth of total cultural pattern'. *The world's population: problems of growth* edited by Q H Stanford (Oxford U P, 1972) contains some useful contributions but is rather disjointed as a collection.

Another expressive title is *Population, resources, environment: issues in human ecology* by P R and A H Ehrlich (second edition, Freeman, 1972), which is described as a 'comprehensive, detailed analysis of the worldwide crisis of overpopulation and the resulting demands on food, resources, and the environment'. Written by two biologists, it ranges broadly over many topics of geographical interest, which are further developed in a second volume entitled *Human ecology: problems and solutions* by P R Ehrlich (Freeman, 1973). The same breadth of approach is seen also in *Population growth and land use* by C Clark (Macmillan, 1968), which touches on biology, agriculture, economics, politics, sociology and history. *World population and world food supplies* by Sir E J Russell (Allen and Unwin, 1954) is a detailed study of a restricted aspect of the problem. The approach is regional. *World population and food supply* by J H Lowry (second edition, Arnold, 1976) is a more recent examination of the same theme.

An historical analysis of population trends is found in *World population: past growth and future trends* by A M Carr-Saunders (Oxford U P, 1936, under the auspices of the Royal Insitute of International Affairs, and subsequently reprinted by Cass). Although this is a general survey of the world situation, the arrangement is topical. *Population in history: essays in historical demography*, edited by D V Glass and D E C Eversley (Arnold, 1965) is a collection of some twenty-seven papers, most of which are either previously unpublished or incorporate substantial revisions from their original form, with emphasis upon Europe and the British Isles. An introductory and very concise global historical study is *The economic history of world population* by C M Cipolla (third edition, Penguin, 1965).

Migrations of population are a fascinating and important field in population geography. A good general study is *Les migrations des peuples: essai sur la mobilité géographique* by M Sorre (Flammarion, 1955), yet another valuable French contribution which is a basic statement and source of reference in this field. A more recently published volume containing a selection of reprinted essays on this theme is *World migration in modern times* edited by F D Scott (Prentice-Hall, 1968). Among numerous regional studies of population movement is *Europe on the move: war and population changes, 1917-47* by E M Kulischer (Columbia U P, 1948), which looks at the demographic effects of two world wars.

Political geography

The literature of politics is described in some detail in two guides. *Political science: a bibliographical guide to the literature* by R B Harmon (Scarecrow Press, 1965, with supplements in 1968, 1972, and 1974) is organised under topics, with each section consisting largely of lists of basic texts and reference sources, and prefaced by a short introduction. A similar arrangement exists in a single-volume guide entitled *The information sources of political science* by F L Holler (American Bibliographic Center-Clio Press, 1971). In this work the emphasis is entirely on reference materials, and in contrast with the previous example each entry is briefly annotated. Bibliographies on aspects of politics are listed in *Political science bibliographies* by R B Harmon (Scarecrow Press), an ongoing project in which two volumes have so far been published (volume one in 1973, and volume two in 1976).

ABC pol sci, published monthly by the American Bibliographic Center-Clio Press (1969-) is a current awareness service which reproduces the contents pages of major journals. At present the periodicals represented regularly amount to about three hundred, and the notification of their contents is often in advance of actual publication. For retrospective searching there is *International political science abstracts* (1951-), a quarterly series published by Blackwell and prepared by the International Political Science Association in cooperation with the International Committee for Social Sciences Documentation. The abstracts are grouped under very broad subjects, and the arrangement is supplemented by a detailed subject index. Useful dictionaries include F Elliott and M Summerskill, *A dictionary of politics* (sixth edition, Penguin, 1970); M Cranston, *A glossary of political terms* (Bodley Head, 1966); and G K Roberts, *A dictionary of political analysis* (Longmans, 1971).

Political geography was defined by Hartshorne as 'the study of the areal differences and similarities in political character as an interrelated part of the total complex of areal differences and similarities'. In common with other systematic divisions of geography it therefore employs both its own unique analytical techniques and theories and, as part of the total spatial synthesis, a strong regional element as well. The basic texts, viewed in the light of these criteria, fall into two groups. Among older examples of major studies which concentrate on the thematic aspects are *Principles of political geography* by H W Weigert et al (Appleton-Century-Crofts, 1957), and *Elements of political geography* by S Van Valkenburg and C L Stotz (second edition, Prentice-Hall, 1954, with numerous later reprints). The former is organised around three factors—spatial, cultural, and economic--and includes numerous regional examples; the latter is addressed to a general readership, intending 'to lay a groundwork for a clear, broad understanding of world problems as a basis for further and more detailed reading in the field'. A short bibliography is appended. Two further substantial American textbooks are worth noting here: *Political geography* by P Buckholts (Ronald Press, 1966) is still serviceable, although a new edition would be welcome; *Modern political geography* by E F Bergman (Brown, 1975) is a comprehensive and up-to-date text, though not particularly readable. Rather more successful studies are *Systematic political geography*, H J De Blij (second edition, Wiley, 1973) and *Political geography* by N J G Pounds (second edition, McGraw-Hill, 1972). The first of these is a well-illustrated and well-documented study, comprising both original analysis and reprints of key papers by various authors. This fact and the bibliographies make this a very valuable source of reference. Pounds' study is a comprehensive introduction, again well-illustrated and incorporating bibliographies with numerous regional examples. *Modern political geography* by R Muir

(Halsted, 1975) is a theoretical work with an emphasis on spatial interaction. It is well-written and concise, and very suitable as a text for an advanced political geography course. Perhaps the best introduction, however, is *Our fragmented world: an introduction to political geography* by W G East and J R V Prescott (Macmillan, 1975), a scholarly and comprehensive textbook at a more elementary level.

There are a large number of works in the second category, arranged largely as regional studies, and these tend to date rather more quickly. Most of these books, however, incorporate some general discussion of the scope and nature of political geography, and in some cases this is the more valuable part of their content. A good example is *World political patterns*, L M Alexander (Rand McNally, 1957), which provides in the introductory chapters a concise review of the development and main themes of political geography. *Geography and politics in a divided world* by S B Cohen (second edition, Methuen, 1974) is a geographical view of mid-twentieth century international politics. Following an introductory section which explains ideas and motives of political geography and its main themes, the author cxamines individual power blocs and spheres of contact and influence. A less traditional approach is applied in *The international politics of regions* by L J Cantori and S L Spiegel (Prentice-Hall, 1970). This is a selection of reprinted papers which emphasise the importance of regional political relationships in international affairs, rather than the influence of the great powers.

It is worth noting several shorter introductory statements. *Geography of world affairs* by J P Cole (fourth edition, Penguin, 1972) is a popular and topical discussion of world problems which makes easy and stimulating reading. The approach is partially thematic, and partically regional. There are two volumes in the Hutchinson university library which are useful here: *Geography behind politics* by A E Moodie (1949), and *The geography of state policies* by J R V Prescott (1968). The former title, although old, is still a very readable account which seeks to explain the relevance of geography to politics; the more recent study centres on the relationship between geographical factors and political decision making. Extensive use is made of quotations from previous authorities, which in association with the bibliographies appended to each chapter provide a review of the literature on some recurrent and significant themes of political geography. J R V Prescott has also written the volume on *Political geography* for Methuen's *Field of geography* series (1972). This is a well-balanced, up-to-date, and concise statement of the current position in the field. The subject content and methodology are described and explained, and then selected major themes are explored in turn. G R Crone's *Background to political geography* (Museum Press, 1967) is not a comprehensive review, but an informal and readable discussion of a 'representative selection of contemporary political problems', as an introduction to the relevance of political geography.

There are a number of useful collections, several of which deserve specific mention. *Essays in political geography*, edited by C A Fisher (Barnes and Noble, 1968), consists of original contributions prepared for the 1964 International Geographical Congress. *Politics and geographic relationships* edited by W A D Jackson and M S Samuels (second edition, Prentice-Hall, 1971) is a selection of reprints by geographers and other social scientists including both general theoretical matter and regional case studies, with a substantial introduction by the editors. Finally, and perhaps the most valuable of these compilations, there is *The structure of political geography*, edited by R E Kasperson and J V Minghi (University of London Press, 1970). Some forty essays covering a wide range of interests are grouped around five basic themes (heritage, structure, process, behaviour, and environment), each of which is prefaced by an extensive editorial introduction. An important feature is the bibliographies incorporated in each section.

The various specialised fields of study within political geography are numerous, and the literature is scattered through many of the social sciences, including sociology, law, economics, and history as well as politics. Traditionally, political geography has been most concerned with the field of international relations; with the territorial basis of the state, with the influence and alignment of countries and power blocs, and with boundary and frontier problems. *Geographical aspects of international relations*, edited by C C Colby (University of Chicago Press, 1938 and reprinted by Kennikat Press, 1970), is a selection of writings by geographers on such themes, and a recent study, *The geography of frontiers and boundaries* by J R V Prescott (new edition, Hutchinson, 1967), is a further example of an important contribution by a geographer. *The significance of territory* by J Gottman (University Press of Virginia, 1973) is a useful exploration of a central theme in political geography, including discussion of a number of pertinent contemporary examples. Most of the more important general texts on international relations, however, are the work of political scientists; there is, for example, *Foundations of international politics* by H H and M Sprout (Van Nostrand, 1962), *Politics among nations: the struggle for power and peace*, H J Morgenthau (fourth edition, A A Knopf, 1967), and *The relations of nations*, F H Hartmann (third edition, Macmillan, 1967). All of these are standard references, but many more similar works of value to the geographer could be quoted. The study of international relations brings in questions of international law, and *International law* by L Oppenheim (Longmans, two volumes) may be used as a source of reference. *The law of the sea: offshore boundaries and zones* edited by L M Alexander (Ohio State U P, 1967) is a symposium devoted to an important aspect of international law and edited by a geographer, and the geopolitical implications here are further explored in *The political geography of the oceans* by J R V Prescott (David and Charles, 1975).

These international aspects of the field have been categorised as 'macro political geography', in contrast to further alternative specialisations such as local government and electoral studies which can be described as 'micro political geography'. Electoral geography is the subject of several important case studies, for instance *The social geography of British elections, 1885-1910*, H Pelling (Macmillan, 1967) and *Géographie des élections françaises de 1890-1951*, F Goguel (Fondation Nationale de Sciences Politiques, 1951), and a bibliography, *The geography of elections: an introductory bibliography* by B R Goodey (Center for the Study of Cultural and Social Change, University of North Dakota, 1968).

The overlapping of political geography with related disciplines in the social sciences is matched by a similar interaction with other geographical fields. *Transportation and politics* by R I Wolfe (Van Nostrand, 1964) illustrates the relevance of themes from economic geography; *The great capitals: an historical geography* by V Cornish (Methuen, 1923) illustrates the relationships between history and politics, and between urban, political, and historical geography; Pelling's work on electoral geography (see above) is described in the title as a 'social geography'. These complicated relationships and interdisciplinary approaches should be kept clearly in mind in the search for more detailed literature. The several concerns of human geography do not constitute separate and readily defined entities, and political geography is no exception.

Economic geography

As in the case of political geography and politics, the economic geographer is faced with the whole complex range of economics literature as a potential source material. Fortunately there are several guides which enable the non-specialist to approach this field logically and to exploit its resources. The best of these is *The use of economics literature*, edited by J Fletcher (Butterworths, 1971). This is a symposium of papers by specialists who between them cover the whole field of economics, citing reference materials and basic texts. *Economics and commerce: the sources of information and their organisation* by A Maltby (Bingley; Hamden (Conn), Archon, 1968) is a more concise survey which is biased towards management and business users, but covers the main sources quite adequately. *How to find out about economics* by S A J Parsons (Pergamon, 1972) is the least satisfactory example, and like its companion volume on geography by C S Minto is based on the Dewey classification. *Reference materials and periodicals in economics* by E L Fundaburk (Scarecrow Press, 1971-) is a potentially useful annotated list which is planned to consist eventually of six volumes covering the whole field. To date only volume one, on agriculture (1971) and volume four, on the automotive, chemical, iron and steel, petroleum and gas industries (1972) are available.

A current awareness service for economics is provided by the *Journal of economic literature,* formerly *Journal of economic abstracts.* This is a quarterly published by the American Economic Association, and contains review articles, lists of new books, copies of title pages of current journals, and an index to recently published articles in a variety of periodicals, including some entries with abstracts. *Economic abstracts* is a Dutch publication which appears fortnightly and provides short abstracts of the contents of some 350 journals. These are French, German or Dutch if the original papers were in these languages, and otherwise in English. The service dates from 1953. An index to English language journal articles back as far as 1886 exists in the *Index of economic journals.* This has been published for the American Economic Association since 1962, each volume covering an earlier period of several years.

Useful aids to interpreting the terminology of economics are *Dictionary of economic terms* by A Gilpin (fourth edition, Butterworths, 1977), and *Dictionary of economics and commerce* by J Hanson (fourth edition, MacDonald and Evans, 1974). Foreign terminology is conveniently equated with English in *A systematic glossary of selected economic and social terms* edited by I Paenson (Pergamon, 1963). Languages included are Spanish, French, and Russian. *A dictionary of natural resources and their principal uses* by N Jackson and P Penn (second edition, Pergamon, 1969) is an alphabetical catalogue of plant and mineral resources with brief descriptions and notes on occurrence and uses. A second reference source which provides rather more information is *Chisholm's handbook of commercial geography* (nineteenth edition by L D Stamp, Longmans, 1975). Originally published in 1889, this large volume has been frequently up dated and is a useful source of reference for some categories of information. Following an introductory section on general factors relating to production and distribution of goods the book falls into two parts: the first is arranged by commodities and provides factual details of each; the second section is regional. While most of the statistical information is out of date and should be supplemented from more frequently published sources, the handbook provides in a convenient form a large quantity of basic data, well organised and indexed for quick reference.

There is an almost embarassing wealth of introductory texts in economic geography. Typically these are large and profusely illustrated, and contrive to review the production, distribution and consumption of commodities on a world scale. This is accomplished by means of an approach which is either directly regional, or commodity orientated, or centred on patterns of economic activity. These compilations date rapidly, and a few examples will suffice; *Economic geography*, C F Jones and G G Darkenwald (third edition, Macmillan, 1965), and *A geography of world economy* by H Boesch (second edition, Van Nostrand, 1974) are typical. More modern approaches are increasingly theoretical. Texts similar to the above but presenting a more accurate impression of the importance of recent theoretical developments in economic geography are *The geography of economic activity* by R S Thoman, E C Conkling and M H Yeates (third edition, McGraw-Hill, 1974); *World economic activities: a geographical analysis* by R M Highsmith and R M Northam (Harcourt Brace, 1969); and *The geography of economic systems* by B J L Berry, E C Conkling, and D M Ray (Prentice Hall, 1975). The last of these is the most successful in achieving the essential balance by mixing theoretical chapters with specific case studies, selected from all parts of the world. The most recent of these attempts to present the theoretical foundations of economic geography, and one of the most readable, is *Location in space: a theoretical approach to economic geography* by P Lloyd and P Dicken (second edition, Harper and Row, 1977), a well-organised and logical introduction at first year undergraduate level.

The shorter introductions to the field may be characterised in a very similar fashion. *The geography of economics: a world survey,* by G Parker (second edition, Longmans, 1972), and *A geography of production* by O Hull (Macmillan, 1969) follow the traditional framework based on commodities and the classification of activity patterns. *A preface to economic geography* by H H McCarty and J B Lindberg (Prentice Hall, 1966) is a most useful introduction to modern concepts and techniques. The scope and methodology are discussed in some detail, and there follows a review of the major subject areas with which economic geography is concerned. *Geography and economics* by M Chisholm (second edition, Bell, 1970) examines the application of the laws and concepts of economics to geographical analysis and stresses the interdependence of ideas from both disciplines. *Economic geography* by B W Hodder and R Lee (Methuen, 1974) is a more comprehensive introductory study, also greatly concerned with the application of economic theory in geography. *The ecology of natural resources* by I G Simmons (Arnold, 1974) is a most unusual book, sufficiently broad in scope to warrant inclusion at this point. This introduces the basic concepts about each resource, discussing man's use of it and the resulting feedback loops in ecological, economic, and social terms.

Readings in economic geography, edited by H G Roepke (Wiley, 1967) is a collection of reprinted papers covering a great variety of topics, including a number of regional case studies. The preface states that the 'selection may contain fewer examples of the use of quantitative techniques than some would prefer'. This deficiency, however, may be satisfied by a second collection, entitled *Readings in economic geography: the location of economic activity*, and edited by R H T Smith, E J Taaffe, and L J King (Rand McNally, 1968). This is an important collection of papers, mainly by geographers, which view economic geography in terms of quantitative locational analysis.

The significance of this approach in modern economic geography is such that the several general texts on location theory can almost be regarded as the most satisfactory introductions to the subject, rather than simply as studies of its methodology. *The economics of location* by A Lösch (Yale U P, 1954, translated from the second German edition of 1944 by W H Woglum) is a classic statement. It includes numerous references and relies heavily on mathematical techniques, and is not easy reading, but remains an essential source.*The location of economic activity* by E M Hoover (McGraw-Hill, 1948) makes easier reading, and is arranged around four central themes: locational patterns and preferences; locational change and adjustment; the locational significance of boundaries; and locational objectives and public policy. More recent important studies include *Location and space-economy* by W Isard (MIT, 1956), and W Alonso, *Location and land use: towards a general theory of land rent* (Harvard U P, 1964). The scope of the former is further clarified by its sub-title, A general theory relating to industrial location, market areas, land use, trade, and urban structure'. The approach of both volumes is mathematical; 'the method is that of economics, and the concern ultimately geographic'. In the series *Lund studies in geography* is *Behaviour and location: foundations for a geographic and dynamic location theory,* A Pred (Royal University of Lund Department of Geography, 1967). This both introduces the concept of theoretical locational analysis and includes a substantial section on real-world deviations from economic location theory. A simple introduction to the technicalities of the quantitative methods used in modern research in this field is found in *An introduction to quantitative analysis in economic geography,* M H Yeates (McGraw-Hill, 1968). A most valuable review of literature on locational theory is to be found in *Distance and human interaction: a review and bibliography*, G Olsson (Regional Science Research Institute, 1965). This discussion ranges widely over examples pertinent to several aspects of human geography, the scope being defined as 'literature in which the distance variable in spatial interaction has been treated'. This may be supplemented by a further volume in the same series, of narrower scope but including many valuable references to pioneering work in this field: *Diffusion processes and location: a conceptual framework and bibliography* by L A Brown (Regional Science Research Institute, 1968).

The same theme is continued in more limited branches of economic geography, and particularly in studies of the geography of industrial activity. *Theory of the location of industries* by A Weber (University of Chicago Press, 1929, translated by C J Friedrich) is a classic in the literature of locational analysis, and although one of the earliest statements on the subject is still an essential work. A more recent study which includes a review of earlier attempts to provide a satisfactory theoretical framework for locational analysis is *Plant location in theory and in practice: the economics of space* by M L Greenhut (University of North Carolina Press, 1956). *Locational analysis for manufacturing,* edited by G J Karaska and D F Bramhall (MIT Press, 1969) is a selection of readings in the *Regional science studies* series, edited by W Isard. It is designed to fill a gap between works of abstract theory and detailed case studies of particular industries; it concentrates on general location factors common to many industries in real life situations. *Spatial perspectives on industrial organisation and decision making*, edited by F E I Hamilton (Wiley, 1974) is another attempt to bridge the same gap. It is a collection of original articles of a high standard, containing a mixture of theoretical papers and case studies. Additional literature is cited in two useful bibliographies. Uniform with Olsson's work (see above) is *Industrial location: a review and annotated bibliography of theoretical, empirical, and case studies* by B H Stevens and C A Brackett (Regional Science Research Institute, 1967). *Industrial location and regional economic policy; a selected bibliography* by P M Townroe is a short and briefly annotated list of mainly British material, published by the University of Birmingham Centre for Urban and Regional Studies, 1968.

Among a number of less theoretical introductory studies of industrial geography are *Geography of manufacturing* by G Alexandersson (Prentice-Hall, 1967) and a rather longer and more comprehensive work with the same title by E W Miller, also published by Prentice-Hall, 1962. The former is a concise introduction to general principles of industrial location and a discussion of selected industries and regions. The latter consists largely of a description of world patterns, partly by region and partly by industry. *A geography of manufacturing* by H R Jarrett (MacDonald and Evans, 1969) is a general introductory textbook which is about evenly balanced between a review of key factors in the choice of location and case studies of selected industries. A similar pattern, although with a lesser emphasis on case studies, is followed in *Industrial activity and economic geography: a study of the forces behind the geographical location of productive activity in manufacturing industry* by R C Estall and R O Buchanan (third edition, Hutchinson, 1973). *Industrial geography* by R C Riley (Chatto and Windus, 1973) is an excellent and concise textbook eminently suitable at undergraduate level. Moving on from these general texts there are numerous examples of the application of the criteria they describe in specialised studies of individual industries. These include such titles as *The geography of iron and steel*, N J G Pounds (second edition, Hutchinson, 1966), *An economic geography of oil*, P R Odell (Bell, 1963), and E M Hoover's classic study of *Location theory and the shoe and leather industries* (Harvard U P, 1937), from which was developed the theoretical framework set out in his general treatise on *The location of economic activity*.

A second major field within economic geography is agriculture, which clearly has links not only in economics but in physical geography as well. In this area, therefore, the bibliographic materials of economics must be used in conjunction with those in agriculture, described in the previous chapter with reference to soils. An additional source, which emphasises both the physical and the human side of agriculture, is *The literature of agricultural research* by J R Blanchard and H Ostvold (University of California Press, 1958). Unfortunately this is now somewhat dated. *Farming systems of the world* by A N Duckham and G B Masefield (Chatto and Windus, 1970) is a textbook on comparative agriculture which provides for the geographer a comprehensive classification and description on a world scale of agricultural regions together with a general analysis of location factors and trends. A selective but very readable collection of case studies from a wide variety of environments is found in *Types of rural economy: studies in world agriculture* by R Dumont (Methuen, 1957, translated from the French edition of 1954).

In addition to the above, there are several more specifically geographical texts. The most comprehensive of these is a two-volume survey entitled simply *Agricultural geography* by P Laut (Nelson, 1968). There are more concise introductory volumes, both sharing the same title as the above, in *Bell's advanced economic geographies* series and in *The field of geography* series respectively. The former, by L Symons (Bell, 1967), is in three parts, which discuss the physical and social environment, the systems by which this is exploited in selected areas, and finally the concepts of regions and land use. The second volume, by W B Morgan and R J C Munton (Methuen, 1971), is characteristic of the series of which it forms a part; it is a brief but authoritative, comprehensive, and modern statement of key concepts in the field, and includes a useful bibliography. Also very valuable as a bibliographic source is the volume on agriculture in Prentice-Hall's *Foundations of economic geography series* (see below). This is by H F Gregor and is entitled *Geography of agriculture: themes in research* (1970), and although there is no bibliography as such the text contains numerous references to further literature cited in footnotes. The most recent of these general introductory texts is J R Tarrant's *Agricultural geography* (David and Charles, 1974), which emphasises methodological themes. *The agricultural systems of the world: an evolutionary approach* by D B Grigg (Cambridge U P, 1974) puts contemporary agriculture into its historical perspective.

The location of agricultural production by E S Dunn (University of Florida Press, 1954) applies mathematical techniques of locational analysis to agriculture, thus emphasising the operation of socio-economic factors. A second and more recent example which analyses mathematically the theoretical basis of rural land use decision making is *A theoretical approach to rural land-use patterns* by W C Founds (new edition, Arnold, 1974). A substantial bibliography is included. By contrast, there is *Ecological crop geography* by K H W Klages (Macmillan, 1942), which analyses the physical requirements of crops and stresses the role of the environment in terms of soils and climate. The same theme is the subject of *Weather and agriculture*, a symposium edited by J A Taylor (Pergamon, 1967). Detailed studies of particular agricultural systems include *A world geography of irrigation*, L M Cantor (Oliver and Boyd, 1967), and *Plantation agriculture*, P P Courtenay (Bell, 1965). The latter is a further example from *Bell's advanced economic geographies*, an important series which also includes an interesting recent regional case study of British farming by J T Coppock, entitled *An agricultural geography of Great Britain* (1971). A specialised and scholarly study which cannot be omitted here is *Agricultural origins and dispersals*, C O Sauer (American Geographical Society, 1952, second edition by MIT Press, 1969). Sub-titled *The domestication of animals and foodstuffs* this is an historical analysis of the evolution and spread of agricultural practices. The original text of the series of Bowman Memorial Lectures of 1952 is supplemented in the second edition by three new papers and a new introduction.

While the geographical analysis of agricultural and industrial activity constitute perhaps the most coherent and clearly recognisable divisions of economic geography and the most prolific in terms of literature, there are numerous other topics in economics which are amenable to geographical treatment. Notable areas of special interest are marketing and distribution, transport, resource utilisation, and trade, international and domestic. Examples from several of the above are included in the *Foundations of economic geography* series, published by Prentice-Hall. This is a series of short texts which are 'designed to bring the student . . . to the frontiers of knowledge in economic geography and . . . to demonstrate the methodological implications of current research . . .' Each volume is written by a specialist and includes either a bibliography or bibliographic footnotes; 'the series as a whole is intended to provide a broad cross-section of on-going research in economic geography'. A second series which provides several relevant examples is the geography section of the Hutchinson university library, certain volumes of which have already been cited in other contexts. On transport there is *The geography of air transport*, K R Sealy (revised edition, 1966); *The geography of sea transport*, A Couper (1972); *Railways and geography* by A C O'Dell and P S Richards (second edition, 1971); and *Seaports and sea terminals*, J Bird (1971). Other titles from the same publisher and in the same series include *The geography of energy* by G Manners (third edition, 1977), and *Geography and retailing* by P Scott (1970). More specialised recent texts are *Geography and inequality* by B E Coates, R J Johnston, and P L Knox (Oxford U P, 1977), a geographical study of a major socio-economic problem; and *Organisation, location, and behaviour: decision making in economic geography* by P Toyne (Macmillan, 1974), a particularly well-written and stimulating textbook.

Economic geography is an important branch of the subject and one which is developing very rapidly. It is also significantly one of the few branches to which major specialised journals are devoted. *Economic geography* (1925-) is a quarterly published by Clark University which interprets the theme fairly broadly and is an essential reference for current research work. Each issue includes several scholarly articles and a number of book reviews. Issues are frequently devoted to special topics, under guest editorship. A second journal of importance is *Tijdschrift voor economische en sociale geographie* (1910-), published by the Royal Dutch Geographical Society (Koninklijk Nederlands Aardrijkskundig Genootschap). As the title indicates, its interests extend beyond economic geography, although there is a high proportion of economic articles. The authorship is international and most contributions are in English, with some in Dutch and some emphasis on Dutch geographical problems in the selection of topics. There are occasional review articles, and each bi-monthly issue contains a number of book reviews.

Settlement geography–rural and urban

Settlement geography is a field which relates closely not only to practically every other aspect of human geography but to a range of outside subjects. Unlike political or economic geography, for example, which both draw from well recognised disciplines, settlement geography has no clear external counterpart; relevant subjects include sociology, economics, demography and politics, as well as more peripheral areas such as architecture and fashionable themes like conservation and the environment. Perhaps the nearest equivalent non-geographical subject is planning. This is a term which can be interpreted with suitable breadth, as is the case in the best literature guide, entitled *Sourcebook of planning information* by B White (Bingley; Hamden (Conn), Linnet, 1971). Only four pages from a total of more than six hundred are explicitly regarded as urban geography, but in fact a much larger proportion of the book contains information of relevance in this branch of geography. Topics covered include the history and present structure of planning activity and discussion of a wide range of source materials, including maps, photographs, reports, and statistics, as well as books, journals, and bibliographic and other reference works. The United Nations Economic Commission for Europe has issued a useful *Directory of national bodies concerned with urban and regional research* (1968) which includes entries for most of the European nations as well as the USA and the USSR.

The most effective abstracting and indexing services are again those in the field of planning and the associated professions of architecture and surveying. The Department of the Environment issues a monthly

Index to periodical articles (1949-), which like its counterpart which lists books–*Classified accessions list* (bi-monthly, 1949-)–is based on the holdings of the department's library, and reflects its interests. Less central from the geographer's point of view are the quarterly *Library bulletin* (1946-) of the Royal Institute of British Architects (and the annual cumulations in the *Annual review of periodical articles*, 1967-), and the *RICS abstracts and reviews* (Royal Institution of Chartered Surveyors, monthly, 1965-). These professional services include much that is irrelevant to the geographer, but usefully supplement the DOE publications in their coverage of the planning literature. *Ekistic index*, produced by the Athens Center of Ekistics, is a monthly computer-produced index serving the 'science of human settlements', and in terms of its subject scope is perhaps the most relevant of this group for geographers. The layout and indexing however do not make it particularly easy to use.

In addition to these continuing publications there are a number of useful, monograph bibliographies which provide a starting point for literature searching on this subject. Two fairly recent examples are *A bibliography of city and regional planning* by K McNamara (Harvard U P, 1969), which is an extensive list of both books and articles including many early contributions, and *Comprehensive urban planning: a selective annotated bibliography with related materials* by M C Branch (Sage Publications, 1970) which stresses American sources. *Central place studies: a bibliography of theory and applications* by B J L Berry and A Pred (Regional Science Research Institute, 1965) is more obviously geographical. Originally published in 1961, the later edition is an unaltered reprint, but includes a supplement listing both omissions from the previous version and new literature up to 1964. Sociological aspects of planning are covered to 1963 by *Land use planning and the social sciences: a selected bibliography*, prepared by the Centre for Urban Studies at University College, London, and published in 1964. A large number of very valuable short bibliographies on planning and associated topics have been published in the USA by the Council of Planning Librarians. These are in mimeographed form, and have been issued at frequent but irregular intervals since 1959. Number 509 in this series, issued in 1974, is an index to topics covered to that date.

Many of the most important journals are those published for professional planners. These include the *Town planning review* (Department of Civic Design, University of Liverpool, 1910-, quarterly); *Journal of the Royal Town Planning Institute* (1914-, ten issues per annum); *Urban studies* (Longmans for the University of Glasgow, 1964-, three issues per annum); *Planning outlook* (Oriel Press for the University of Newcastle upon Tyne, 1948-, twice yearly); and the more popular, less academic monthly *Town and country planning* (Town and Country Planning Association, 1932-). The principal American contribution is the *Journal of the American Institute of Planners* (1917-, bi-monthly). Of broader scope but including much relevant material are periodicals such as the *Journal of regional science*, which has been described previously (chapter four), and *Regional studies*, published quarterly by Pergamon for the Regional Studies Association and under geographical editorship. *Ekistics* (1955-) is a monthly journal from the same source as *Ekistic index*, and like the latter interprets the field very loosely, ranging widely over scientific and technological as well as social and human issues.

A good short introduction to basic concepts in settlement geography is *Settlement patterns* by J A Everson and B P FitzGerald (Longmans, 1969). A stated objective of the authors is the explanation in simple terms of modern theoretical concepts, and this has been achieved, from the pioneering work of Walter Christaller onwards. Christaller's own study on southern Germany is a classic in this field and is available in English as *Central places in southern Germany* (Prentice-Hall, 1966), translated by C W Baskin. The original German edition was published in 1933. There are several alternative texts which introduce concisely the narrower field of urban geography; these include *The geography of towns*, A E Smailes (revised edition, Hutchinson, 1966); *Towns and cities,* E Jones (Oxford U P, 1966); and *Urban geography: an introductory analysis* by J H Johnson (second edition, Pergamon, 1972). This last mentioned is the best short general introduction and includes well chosen lists of supplementary readings. *Inside the city* by J A Everson and B P FitzGerald (Longmans, 1972) is a more elementary text uniform with *Settlement patterns* cited above.

More substantial basic textbooks include examples by British, American and French geographers. The most recent British work is *The study of urban geography* by H Carter (second edition, Arnold, 1976). This is more advanced than the above books and attempts to cover the subject to degree level, indicating the direction of current research on each of the several main themes of urban geography. *Urban and regional geography* by P Hall (David and Charles, 1975) applies geographical principles in a practical context, and although by a geographer is written with the layman very much in mind. A major contribution with an emphasis on the external physical, social, and economic relationships of urban centres with their surrounding regions is *City and region: a geographical interpretation* by R E Dickinson (Routledge and Kegan Paul, 1964). *Traité de géographie urbaine* by J Beaujeu-Garnier and G Chabot, published in 1953, is available in an English translation by G M Yglesias and S H Beaver (Longmans, 1967). This again is an advanced general treatment with examples from all over the world and a useful bibliography which is particularly strong on French sources. The American example is *Geographic perspectives on urban systems, with integrated readings* by B J L Berry and F E Horton (Prentice-Hall, 1970). This is a compilation and synthesis of a great number of reprints of previously published papers, which have been skilfully edited and integrated to create a most useful general study of urban geography. Unlike many similar works the selections are well introduced and given a place within a coherent and logical framework.

A much older but still valuable collection is *Readings in urban geography*, edited by H M Meyer and C F Kohn (University of Chicago Press, 1959). This consists of over fifty papers, mainly by geographers but including some by other social scientists covering most phases of urban geography. *The study of urbanization*, edited by P M Hauser and L F Schnore (Wiley, 1965), is a consciously interdisciplinary collection, which both reviews the status of urbanisation as a research theme in various social sciences and comments on selected research problems. Extensive and very useful lists of references accompany each paper. A third reader is *A geography of urban places*, edited by R G Putnam, F J Taylor, and P G Kettle (Methuen, 1970), in which some forty essays are arranged under three themes: firstly the definition of urban geography, 'the origin, location, and characteristics of cities'; secondly, the economic basis and function of cities; and thirdly, the effects of urbanisation, social and environmental. In addition to these collections of reprinted articles, mention should be made of two important symposia. The first of these deals with a specialised aspect of urban geography, under the title *Urban core and inner city*. This is the proceedings of a conference held in 1966 and organised by the Sociographical Department of the University of Amsterdam. The second, which includes some important theoretical papers, is the *Symposium on urban geography* held at Lund in 1960, under the auspices of the International Geographical Union. The proceedings, edited by K Morborg, are published in *Lund studies in geography*.

Many significant contributions to the literature of urban geography take the form of case studies, either of particular cities or of city development within certain areas. Important examples of the latter are *The West European city: a geographical interpretation* by R E Dickinson (second edition, Routledge and Kegan Paul, 1961); *The American city: an urban geography*, R E Murphy (second edition, McGraw-Hill, 1974); *The North American city*, M Yeates and B J Garner (second edition, Harper and Row, 1976); *The containment of urban England* edited by P Hall (two volumes, Allen and Unwin, 1973); and *The towns of Wales*, H Carter (second edition, University of Wales Press, 1966). A major regional work with an historical approach is *International history of city development*, E A Gutkind (Collier-Macmillan, 1964-); This is a series which 'is planned to include eventually the whole world: to date eight volumes have appeared on different parts of Europe. The complete set will be an impressive reference source. A selection of important cities are described in *The world cities* by P Hall (Weidenfeld and Nicolson, 1966) and *Great cities of the world: their government, politics and planning* edited by W A Robson and D E Regan (third edition, Allen and Unwin, 1972). *Urban analysis: a study of city structure with special reference to Sunderland* by B T Robson (Cambridge U P, 1969) provides a local illustration of the methodology and theory of modern urban geography, based on a single city.

The volume of literature devoted to rural as distinct from urban settlement (a distinction which is not in fact very clear in many cases) is relatively small. A good concise account is found in *Rural settlement and land use: an essay in location* by M Chisholm (Hutchinson, 1962), which includes a short and highly selective bibliography, and more recently in *Rural geography: an introductory survey* by H D Clout (Pergamon, 1972). The latter is not as comprehensive as its title suggests, but is the best modern survey available in English. Naturally a high proportion of the available sources tend to concern themselves as much with agriculture and general problems of rural land use as with settlement, the two themes being inextricably linked. The whole field is covered by the term 'rural geography', as used by P George in his *Précis de géographie rurale* (Presses Universitaires de France, 1967). A second field in which many relevant works are found is rural sociology. A useful example, consisting partly of general essays and partly of regional case studies from various European countries, is *People in the countryside: studies in rural social development*, edited by J Higgs (National Council of Social Service, 1966). A larger and more general collection is *Rural sociology: an analysis of contemporary rural life*, edited by A L Bertrand et al (McGraw-Hill, 1958), which is designed as a broad introduction to the field. *People and the countryside* by H E Bracey (Routledge and Kegan Paul, 1970) is a well-documented volume based on the British experience which provides a thoughtful and stimulating analysis of current rural problems. *A systematic source book in rural sociology* edited by P A Sorokin, C C Zimmerman, and C J Galpin (University of Minnesota, 1930-2, and reprinted by Russell and Russell, 1965) is a three volume reference source, which includes a range of readings, with editorial comment and additional bibliographies. *Sociologia ruralis* is the most useful journal in this field. This is the journal of the European Society for Rural Sociology, published quarterly and including articles in English, French and German. A recent analysis of the content of the journal shows rural settlement to be the subject of the majority of papers published (Volume XII, no 1, 1972, 86-103).

Historical geography

The field of historical geography represents probably the most extreme case of the dependence of a specialised branch of geography on the literature of another discipline. Firstly, although some aspects of history are of much greater relevance than others—economic history, for example, is obviously more amenable to geographical study than is constitutional history—the geographer requires access to an enormous range of historical sources. Secondly, the literature of historical geography, as distinct from history, is relatively small. In particular there is a shortage of general texts introducing the concept of historical geography, defining its scope and character, and explaining its methodology. To a large extent this has to be achieved by deduction from case studies. At the same time, the journal literature is dispersed in a great variety of sources, historical, geographical and archaeological; this pattern is certain to persist in spite of the recent foundation of a specialised *Journal of historical geography* (Academic Press, quarterly, 1975-).

The most useful general statement, apart from a number of scattered papers in journals, is *Historical geography* by J B Mitchell (English Universities Press, 1954). The continuing validity of this small work is emphasised by the fact that it is still in print after more than twenty years. It is still adequate as a guide to major concepts, techniques, and themes, but can be usefully supplemented by *Progress in historical geography* edited by A R H Baker (David and Charles, 1972), a collection which describes more recent work in the field. A second volume of essays which defines the essential nature of the field in its title is *Geographical interpretations of historical sources: readings in historical geography*, edited by A R H Baker, J D Hamshere, and J Langton (David and Charles, 1970). This consists of a set of reprinted papers which cover a range of topics in the historical geography of England and Wales, with an introductory essay by the editors. Some of the papers contain amendments and supplementary notes in addition to their original text. A R H Baker has also edited (this time with J B Harley) an interesting and well-illustrated collection of reprints from the *Geographical journal*, under the title *Man made the land: essays in English historical geography* (David and Charles, 1973). A similar function in respect of the United States is served by *Pattern and process: research in historical geography* edited by R E Ehrenburg (Harvard U P, 1975), a collection of nearly twenty papers illustrating American work in historical geography. *An historical geography of England before AD 1800*, edited by H C Darby (Cambridge U P, 1936) is another collection of specialised papers which serves as a good illustration of the scope of historical geography, and has become one of the classics in this field. More recently Darby has edited a second important collection entitled *A new historical geography of England* (Cambridge U P, 1973), most of which is concerned with the nineteenth century. The similarity is somewhat superficial, however, as the 'snapshot' technique of the former work (examining the geography of successive periods in turn) is abandoned in the *New historical geography* in favour of a more evolutionary and thematic approach. Geographical history, not historical geography, is the subject of *The geography behind history* by W G East (revised edition, Nelson, 1965) and *A geographical introduction to history* by L P V Febvre and L Bataillon (Barnes and Noble, 1966, translated from the French by E G Mountford and J H Paxton).

With a few exceptions such as the above, the literature of historical geography consists principally of regional works. For the United States there exists a useful guide to these in *Historical geography of the United States: a bibliography* by D R McManis (Eastern Michigan University, 1965). This lists some 3500 items from journals and books, grouped under states. Europe is covered by two textbooks on historical geography, both of which include lists of references to further sources: *An historical geography of Europe* by W G East (fifth edition, Methuen, 1966), and *An historical geography of Western Europe before 1800* by C T Smith (Longmans, 1967). A further source for the early period, which however excludes the USSR and the British Isles, is *An historical geography of Europe, 450 BC–AD 1330* by N J G Pounds (Cambridge U P, 1973), which looks at five successive periods. Russia is covered in *An historical geography of Russia*, W H Parker (University of London Press, 1968), and France in a recent collection, *Themes in the historical geography of France*, edited by H D Clout (Academic Press, 1977). A major series is *The Domesday geography of England* edited by H C Darby (Cambridge U P, 1954-). Several volumes have already appeared, and more are planned, each devoted to a particular region. This is a classic study in the methodology of historical geography, recreating regional geography at a specific point in time, using historical sources. *Studies of field systems in the British Isles*, edited by A R H Baker and R A Butlin (Cambridge U P, 1973) consists of twelve chapters by specialists on field systems in different areas, with introduction and conclusion by the editors and an excellent bibliography.

There are a large number of bibliographical guides to historical literature, including several which are international in scope. Among the latter are two annual services, the *Annual bulletin of historical literature* and the *International bibliography of historical sciences*, which have been published since 1911 and 1926 respectively. The former is prepared by the Historical Association, and each issue consists of brief critical reviews, written by specialists, of selected publications of the year arranged principally by period. The *International bibliography* is edited for the International Committee of Historical Sciences in English, French and German, and is a select, classified list. *Historical abstracts* (American Bibliographic Center-Clio Press, 1955-) is also international in its coverage. Previously confined to the period 1775-1945, it is now published in two parts, on modern history (1775-1914) and the twentieth century (1914-). Both series are arranged mainly regionally and both are published quarterly, the last issue of each volume consisting of a cumulative index. *Archäologische Bibliographie* (de Gruyter, 1913-) is an annual bibliography of archaeological literature based partly on regional and partly on thematic headings.

Among many monograph bibliographies a useful example of fairly general application, and again of international scope is *A bibliography of modern history*, edited by J Roach (Cambridge U P, 1968). This consists of collected bibliographies for the *New Cambridge modern history:* the predecessor of this standard work contained separate bibliographies in its various volumes. Of all branches of history the closest to historical geography is economic history: *British economic and social history: a bibliographical guide* by W H Chaloner and R C Richardson (Manchester U P, 1976) contains a selection of well over four thousand references arranged under subjects within four broad periods. The literature of history is to a large extent regional in character, and as in the case of regional geography the scale of study differs widely, from the parochial to the global. The general sources indicated above have clearly to be supplemented by more detailed local materials; useful illustrations on a national scale include *British archaeological abstracts* and *Bibliographie annuelle de l'histoire de la France.* While it is beyond the scope of this volume to pursue this aspect in detail, a valuable source for tracing such works should be noted. This is the American Historical Association's *Guide to historical literature* (Macmillan, 1961) which lists bibliographies, reference works, major texts, biographies, government publications, and other materials. It is divided regionally and by period, and the coverage is worldwide. There is also an introductory guide of a more general nature in *How to find out in history* by P Hepworth (Pergamon, 1966), but this is far less comprehensive. There are also two guides similar in style to C B M Lock's *Geography and cartography: a reference handbook*; both are by A E Day, and are *History: a reference handbook* (Bingley; Hamden (Conn), Linnet, 1977), and *Archaeology: a reference handbook* (Bingley; Hamden (Conn), Linnet, 1977), containing annotated entries on key topics, themes, and sources. *The parish chest: a study of the records of parochial administration in England* by W E Tate (third edition, Cambridge U P, 1969) is a guide to the unpublished records of English local history which is a most useful companion for the detailed study of particular areas. It may be supplemented by *Sources for English local history* by W B Stephens (Manchester U P, 1972). An American example is the *Harvard guide to American history*, edited by O Handlin et al (second edition, Harvard U P, 1974).

Miscellaneous special topics

As previously stated, the classification of the major divisions of human geography is not easy, and there are a wide variety of specialised themes which have been studied geographically. Those discussed below are indicative of this trend rather than a comprehensive catalogue. Although in most cases these themes can be associated primarily with one or other of the major branches detailed above, they are each also related in a more complex way to several of these branches, and for this reason they are included in a separate section.

Medical geography is a field which has gained considerably in popularity in recent years. There are two good short introductions by L D Stamp, the more comprehensive of which is *The geography of life and death* (Collins, 1964). The other work is the text of a series of lectures given at the London School of Hygiene and Tropical Medicine, under the title *Some aspects of medical geography* (Oxford U P, 1964). Both include brief notes on sources. *Medical geography: techniques and field studies*, edited by N D McGlashan (Methuen, 1972), is a selection of reprints and original papers containing both general material and case studies illustrative of general themes and techniques. Each paper cites further references and one, by A T A Learmonth, reviews atlases in medical geography over the period 1950-70. This volume provides a most useful introduction to the subject. *Environmental medicine*, edited by G M Howe and J A Lorraine (Heinemann, 1972) is a collection of specialist papers, not a comprehensive review as the title seems to imply. *Man, environment, and disease: a medical geography of Britain through the ages* by G M Howe (David and Charles, 1972) is a useful extended case study by one of the leading exponents in this field. There are several valuable contributions edited by J M May and published in a series entitled *Studies in medical geography*, prepared under the auspices of the American Geographical Society. Most of the topics covered in these volumes to date are related to the ecology of particular diseases, or to topics such as malnutrition within individual countries.

The *Foundations of cultural geography* series, from Prentice-Hall, includes several titles on themes of an interdisciplinary nature which cannot be easily classified. There is, for example, *Geography of domestication* by E Isaac (1970), and *House form and culture* by A Rapoport (1969). A further title in this series, which serves to introduce another field in which several geographical studies have been made, is *Geography of religions*, D E Sopher (1967). This is a good general introduction which includes references to earlier literature, although this is not an area of prolific writing. A recent British study is *The geography of religion in England* by J D Gay (Duckworth, 1971). The French have also made some notable contributions; there is, for example, a general work, *Géographie et réligions* by P Deffontaines (Gallimard, 1948), and an analysis of *Les fondements géographiques de l'histoire de l'Islam* by X de Planhol (Flammarion, 1968).

Often treated as a branch of political geography is military geography. As in most of these relatively narrow fields, the bulk of the literature takes the form of journal articles rather than books. A useful guide to both categories is a mimeographed list entitled *Bibliography of military geography*, compiled by L C Peltier, and published by the Association of American Geographers, 1963. This contains about one thousand references, and provides a starting point for a search for more recent literature. There is also a concise and basic textbook which outlines the scope and objectives of the field. This is *Military geography* by L C Peltier and G E Pearcy (Van Nostrand, 1966).

Recreational geography is a topic which has derived considerable popularity recently from current interest in planning and environmental problems. As with many of these other highly specialised themes the literature is diverse, but there is a useful recent synthesis in *The geography of recreation and leisure* by I Cosgrove and R Jackson (Hutchinson, 1972), and a valuable bibliography is available in *Economics of outdoor recreation* by M Clawson and J Knetsch (Johns Hopkins Press, 1966). *Recreational geography*, edited by P Lavery (David and Charles, 1974), is another collection, mainly written by geographers, which provides a good introduction to the main themes. *Recreation in the countryside: a spatial analysis* by J T Coppock and B S Duffield (Macmillan, 1975) points out problems rather than suggesting solutions. *Land and leisure in England and Wales* by J A Patmore (David and Charles, 1970) is a well-illustrated and balanced account of the provision and demand for recreational areas in Britain. *A geography of tourism* by H Robinson (MacDonald and Evans, 1976) provides a good introductory study of a much neglected and little-documented aspect of recreational geography.

Thirteen

REGIONAL GEOGRAPHY

WHILE MOST of the systematic divisions of geography overlap extensively into other disciplines, and research at the frontiers of the subject may scarcely be recognisable as geographical, there is little dispute about the centrality of regional concepts. Regional geography has traditionally formed part of the core of the subject, and in a very real sense reflects its essential character. To many non-geographers, geography *is* regional geography. In view of this it is remarkable that there are few balanced and reasoned monograph studies of the philosophy, scope, and methodology of regional geography. A readable and systematic discussion is however available in *Regional geography: theory and practice* by R Minshull (Hutchinson, 1967). This is both analytical and historical, and provides a clear understanding of the status of regional geography in the context of geography as a whole. In recent years the trend towards quantification in the analysis of the spatial distribution of phenomena has led to the use of the term 'regional science', which represents a new approach to the traditional regional theme in geography. This view is discussed in *Regional ecology: the study of man's environment* by R E Dickinson (Wiley, 1970). The book falls into three sections, on history, theory, and practice, and explores the meaning and relevance of regionalism in modern geography. Although not derived from traditional interpretations of the scope and nature of regional geography, there are several useful studies of 'regional science' which share the same philosophical basis if not the methodology of area study in geography. In spite of the differences these can throw light on common problems and themes. Certain of these are discussed in chapter ten in terms of quantitative methods; *Introduction to regional science* by W Isard (Prentice-Hall, 1975) is a further example, more broadly based than most, which provides a systematic general statement.

Inevitably, there is a very extensive literature both of regional geographies and of sources for the study of regional geography. The description of this literature poses two significant problems. Firstly, the distinction between systematic and regional geography is, in the context of the literature, an artificial one. Many systematic texts are based on the example of a particular regional case study, and these obviously become the source materials for more comprehensive analysis of a region. The second difficulty is one of scale. The term 'regional geography' implies nothing about the areal scope of local studies, which may be continental or parochial. For these two reasons the potential volume of literature on which this chapter is based is enormous, and some explanation of the following selection is essential.

There are substantial lists of regional texts available in several of the general geographical bibliographies, and newly published works are also traceable through general sources. Major regional studies naturally tend to date very quickly (particularly in terms of human geography) and must be read with caution; some of these go out-of-print quite quickly, but others, suprisingly, remain available long after a large part of their content has become history. They can still be useful, however, provided that the date of publication is carefully noted, and that they are supplemented by more recent material, particularly from the journals. Fortunately it is relatively easy to use the general geographical indexing and abstracting services for identifying materials for regional research. The coverage in these sources of regional literature is typically more comprehensive than that of systematic topics, and both the classification and indexing in most of them is best suited to the retrieval of regional information. This is partly the result of the traditional importance of regional geography and partly because the indexing of named places and localities presents fewer terminological problems than that of more abstract concepts. For these reasons the approach to the literature adopted in this chapter is based on a review of basic general texts and a limited number of the more useful reference sources with an emphasis on recently published works and continuing current publications. There is no attempt at coverage in depth or at treatment on a small scale.

Broadly speaking the discussion which follows falls into two parts. The first deals with general sources of information, describing categories and kinds of material which are available for many countries or areas and suggesting the means of locating local examples. Also included are specific details of some useful reference works and textbooks covering the world as a whole. The second part of the chapter takes the major regions in turn and lists the most useful texts and sources relating to each.

General regional sources

Several categories of literature which are applicable in regional geography concerning any part of the world may be conveniently grouped at this stage and described in general terms. Each of these has been discussed previously in the broader context of general geography, but some reiteration with special reference to regional information is in order.

There are numerous relevant journal titles, which can usefully be classified into several types. Probably the largest group are those produced by local geographical societies, which typically contain an implicit specialisation in matters of local concern. This source may be exploited by the use of the *International list of geographical serials*, which is organised on a regional basis. A second and related category consists of certain non-geographical but locally-oriented serial publications prepared by institutions with particular regional connections; the most notable example of interest to geographers are the *Bank reviews*. The banks of many countries publish journals which are a valuable and up-to-date source of economic data, trends, and conditions in their own sphere of interest. In addition, some of the larger banks with extensive international connections have published information on a more generous scale: Barclays, for example, publish a journal called *Overseas survey*, as well as pamphlets and factsheets on individual countries.

Thirdly, there are the publications of research organisations, associations, and institutions. Many such organisations devoted to the study of particular countries and regions publish journals as a medium for their research interests, and these may be important sources for the geographer. Appropriate examples are quoted below in each section. In a category of its own is the journal *Focus*, published monthly by the American Geographical Society. Each issue comprises a brief survey of an area or country, presenting the most recent information and providing both a useful summary of the current situation and a supplement to more detailed textbook studies.

Reference materials which are of prime importance in regional geography include examples from all of the classes described under general geography. Local societies and institutions are important not only as publishers but in terms of their libraries and of the accumulated experience and regional knowledge of their members. This emphasises the role of directories as a key to unpublished information which may be available in response to a direct enquiry once the source has been located. Locally published materials may not be generally known outside the country of origin, but are listed in the respective national bibliographic services, which are detailed in the standard bibliographies of bibliographies and literature guides. Statistical data can be derived either from international sources or from national statistical yearbooks and other similar local publications. Finally, there is the information available in cartographic form in national, regional, or local atlases and map series.

Apart from these broad categories of sources, which were discussed in greater detail in chapters three to eight, there are a number of individual reference books which provide concisely basic regional factual information on a world scale. The best of these is *Worldmark encyclopedia of the nations* (fifth edition, Worldmark Press, 1978). This is a five volume encyclopedia which attempts to provide uniform data for all the countries of the world, as well as information about the United Nations and its various agencies and associated organisations. A wide selection of factual details is included, covering physical and human geography and including bibliographies. Shorter single volume works of a similar type include the *Larousse encyclopedia of world geography*, adapted from the *Géographie universelle Larousse* and edited by P Deffontaines (second edition, Hamlyn, 1967) and *Cowles encyclopedia of nations* (Cowles Education Corporation, 1968). Both are well-illustrated and the factual content is reliable, but no bibliographies are provided in the former, and references in the latter are minimal. A country-by-country survey, containing geographical, historical, economic and cultural data as well as tourist information is the two volume *Encyclopedia of world travel*, revised by M Zelko (Doubleday, 1967). *A handbook of new nations* by G E Pearcy and E A Stoneman (Van Nostrand, 1968) provides brief descriptions of the main features of selected countries which gained their independence in the 1960s. Generally, these are areas of very rapid change and the nations covered are no longer 'new', but the book still contains materials of interest. The *Europa yearbook*, to which reference was made previously as a directory, is also appropriate here and provides basic factual information on a country-by-country basis, updated annually. More detailed data is available in several other yearbooks from the same publisher covering major regions; these are described below in their respective sections.

The above are all designed specifically as reference works and are organised to facilitate quick and easy consultation. In addition there are several textbooks on world regional geography which can be used for this purpose. These include *World geography*, edited by J W Morris and O W Freeman (third edition, McGraw-Hill,1972); *Regional geography of the world: an introductory survey* by J H Wheeler, J T Kostbade, and R S Thoman (third edition, Holt, Rinehart, and Winston, 1975); *World regional geography* by O H Heintzelman and R M Highsmith (fourth edition, Prentice-Hall, 1973); *The world: a general regional geography* by J H Stembridge (third edition, revised by D Parnell, Oxford U P, 1974); and *Culture worlds* by R J Russell, F B Kniffen, and E L Pruitt (second edition, Macmillan, 1969). All of these are well-illustrated American texts which succeed in supplying basic descriptions of the various regions of the world. They are not strong on references to more specialised literature, and do not therefore make particularly good introductions to detailed regional study. They do have a function, however, as supplements to the alphabetically arranged reference compendia. *The world in focus* by W MacQuity (Bartholomew, 1974) is a collection of spectacular colour photographs showing landscapes from all over the world, arranged in regional sections. Older sources which are still useful include the *Géographie universelle* (see pages 137-8), and the *Geographical handbook* series produced by the Naval Intelligence Division of the Admiralty between 1941 and 1945. These were prepared for some thirty countries and areas, mainly in Europe, North Africa and the Middle East, and published in nearly sixty volumes. In spite of their age and the bias implied by the special reasons for their compilation, these remain of value particularly for studies in regional physical geography.

In addition to the bibliographies on general geography referred to in previous sections and which are obviously relevant to regional geography, there are several which have not been described elsewhere and which should be mentioned here. Each of these covers a broad area spanning two or more of the divisions used below. *The developing nations: a guide to information sources concerning their economic, political, technical and social problems* by E G ReQua and J Statham (Gale Research, 1965) contains a somewhat miscellaneous selection but nevertheless lists many useful items. It includes lists of agencies and institutions, periodicals, directories, bibliographies and texts, and is well annotated. Particularly well covered is the field of foreign aid programmes and technical assistance. Some related themes are explored in *The process of modernization: an annotated bibliography on the sociocultural aspects of development* by J Brodie (Harvard U P, 1969). This is a bibliography of books and articles, classified under the headings of industrialisation, urbanisation, and rural modernisation with various sub-categories. An index of places is incorporated, which reveals a worldwide coverage and which enables the geographer to locate relevant entries readily. *Regional economic analysis and the Commonwealth: a bibliographic guide* by F E I Hamilton (Weidenfeld and Nicolson, 1969) is areally limited as indicated in the title. The lists of references on each country are not annotated but the sections are prefaced by some general comment. The bibliography totals over five thousand items, and the first section (which in contrast to the remainder contains detailed annotations) contains references on the concept of the region, regionalism, regional science and regional planning. *A reader's guide to contemporary history*, edited by B Krikler and W Laquer (Weidenfeld and Nicolson, 1972) is a country by country review of the literature, mainly in English, on current affairs. Political, social and economic themes are usefully covered, and the regional arrangement makes this a convenient guide to basic sources for the geographer in these fields. There is a new series of promising regional bibliographies to be published by the Clio Press. Entitled the *World bibliographical series*, each volume covers one country and is designed 'to express its culture, its place in the world, and the qualities and background that make it unique'. So far only one volume has actually be published, *Yugoslavia* by J J Horton (1977), but others are in preparation, and this could become a useful set for reference purposes.

In addition to the many textbooks on specific countries and regions which are cited below, it is worth noting several valuable regional series. Generally speaking, the titles in each of these conform to a similar format and are written at the same level and for the same market. *The world's landscapes*, edited by J M Houston and published by Longmans, is an undergraduate series of regional geographies based on the theme of landscape change brought about by human activity. (Longmans of course also publish many other excellent regional texts in more general series, notably in *Geographies for advanced study.*) Prentice Hall publish a *Foundations of regional geography* series similar to their other 'Foundations' series in cultural geography and economic geography referred to in the previous chapter. These are concise, authoritative, introductory textbooks written by experts. *Pergamon regional geographies* are rather more variable in both format and quality.

Apart from these basically academic series, there are two additional series of more general interest which include much useful material for the geographer. *Nations of the modern world*, published by Benn, consists of nearly thirty individual studies of a fairly popular nature, some of which are written by established academic geographers. Also intended for a popular market are volumes in David and Charles' *Islands series*, but 'subjects such as geology and history are covered in detail and a close look is taken at the people, economy, livelihood, and communications of each island, bearing in mind the difficulties of situation and climate'. Nearly fifty islands are described, ranging from Singapore to the Hebrides. The simplest method of tracing items in any of these series is by use of the publishers' own catalogues.

Europe

The documentation of European geography is both elaborate and comprehensive in comparison with most of the other continents. Major abstracting or indexing services with an international coverage but also a local bias are produced in France, Germany, Russia and the United Kingdom. In addition most of the European nations have strong national geographical societies which publish journals containing bibliographic information. Thus although there are a great many sources, the majority of these are national rather than continental, and bibliographic research is restricted by this disparate structure. There is no single convenient guide to information sources on a continental scale, although there are some useful guides to individual countries: typical examples, both published by Pergamon, are *How to find out about Italy* by F S Stych (1970), and *How to find out about France* by J E Pemberton (1966).

Almost all of the standard textbooks dealing with Europe as a whole are revised editions of fairly long established works. Perhaps the oldest example which is still of use is *Europe: a regional geography* by M R Shackleton (seventh edition, Longmans, 1964). This was originally published in 1934 and has been almost completely rewritten in subsequent editions; apart from the obvious requirement of up-dating factual information, the original emphasis on the physical landscape has been superseded by a more balanced approach. Like Shackleton, most of the other texts are arranged under regions and countries. *A geography of Europe* by J Gottman (fourth edition, Holt, Rinehart and Winston, 1969) is no exception. This again is at first sight rather dated, but the preface points out that 'in dealing with this rapidly changing part of the world, any attempt at precision and timeliness would be futile; instead, the aim has been to strike a balance between the latest data and the permanent factors, thereby giving the student as broad and deep a view as possible of the present status and general trends of European geography'. The result is a text particularly rich in historical and physical detail. The same feature is apparent in *A geography of Europe*, edited by G W Hoffman (third edition, Methuen, 1969), a collection of essays by specialists covering the major divisions of the continent in turn. Each paper follows a similar pattern, describing the physical landscape, the cultural and historical background, and the political divisions which comprise each region. There are in addition introductory chapters on the continent as a whole, and the collection is well edited to form an integrated text. *Europe and the Soviet Union* by N J G Pounds (second edition, McGraw-Hill, 1966) is an orthodox country-by-country account covering the whole continent briefly. Originally published in 1953 as *Europe and the Mediterranean*, the scope has been extended to include a fuller account of the Soviet Union. Each section concludes with a short bibliography. A refreshingly different approach is followed in *Europe and its borderlands*, A G Ogilvie (Nelson, 1957), which is largely systematic. In contrast to most of the above regionally-organised texts, this provides a useful synthesis of the dominant features of the continent as a whole. *The European culture area: a systematic geography* by T Jordan (Harper and Row, 1973) also employs a general continental approach and is another interesting and distinctive study.

A regional geography of Western Europe by F J Monkhouse (fourth edition, Longmans, 1974) is a comprehensive and detailed text, although the scope of Western Europe is somewhat narrowly defined as France and the Benelux countries. A more satisfactory definition of the area is used in Monkhouse's *The countries of North western Europe* (third edition, Longmans, 1974) and in a volume by A F A Mutton which is very much less detailed and includes brief accounts of all countries west of the iron curtain. This work, *Western Europe in colour: an advanced regional geography* (Blandford, 1971) relies heavily on its many excellent maps, which form an essential part of the whole. *Regional development in Western Europe*, edited by H D Clout (Wiley, 1975) is not a systematic survey but a series of essays on selected national problems.

Western Europe is an area with a long tradition of geographical writing and there are in addition to the above general works a great many more regional studies on a smaller scale. *France: a geographical survey* by P Pinchemel (Bell, 1969, translated from the second French edition by C Trollope and A J Hunt) is a still-useful example by a representative of a national school of geography which has traditionally placed great emphasis on regional study. A second example also available in translation is *France: a geographical study* by P George (Martin Robertson, 1973); this is translated from the French edition of 1967 (by I B Thompson) but is up-dated with footnotes and statistics. *France* by J Beaujeu-Garnier (Longmans, 1976) is less readable in translation and could do with better illustration, but is another sound basic text. A good alternative by a British geographer is *Modern France: a social and economic geography* by I B Thompson (Butterworth, 1970). Germany is less well-covered in English, but a good introductory survey is provided in *Germany: its geography and growth*, K A Sinnhuber (second edition, Murray, 1970), which is concise and lavishly illustrated. *Germany*, T H Elkins (second edition, Chatto and Windus, 1968) is another short but useful and well-written study. Both volumes include both East and West Germany, although the emphasis is understandably on the west. A more recent study is *West Germany: an introduction* by G Kloss (Macmillan, 1976); by a linguist rather than a geographer, this is a useful survey of the main features of contemporary Germany. The Benelux countries are covered in a concise and readable text, *Benelux: an economic geography of Belgium, the Netherlands and Luxembourg* by R C Riley and G J Ashworth (Chatto and Windus, 1975). There are several basic books covering the Scandinavian countries: *A geography of Norden: Denmark, Finland, Iceland, Norway, Sweden*, edited by A C Z Sømme (new edition, Heinemann, 1968), a volume originally prepared for the 1960 International Geographical Congress in Sweden, is still one of the best. A good straightforward and more up to date study is *Scandinavia* by B Fullerton and A F Williams (second edition, Chatto and Windus, 1975). This includes a bibliography, mostly of English language sources.

The best standard general geographies of Britain are *The British Isles: a geographic and economic survey*, L D Stamp and S H Beaver (sixth edition, Longmans, 1971), and *The British Isles: a systematic and regional geography* by G H Dury (fifth edition, Heinemann, 1971). Both of these have been established texts for several years and both are substantial and detailed studies; Dury's work contains an extensive bibliography. These textbooks are usefully supplemented by two important collections: *The British Isles: a systematic geography*, edited by J W Watson and J B Sissons (Nelson, 1964), and *Great Britain: geographical essays*, edited by J Mitchell (English Universities Press, 1962). Both of these are still in print and still of use, in spite of their age.

Other areas of Europe are much less well documented, at least in terms of sources in English. *Southern Europe: the Mediterranean and Alpine lands* by R and M Beckinsale (University of London Press, 1975) is a broad survey which largely supersedes M Newbigin's classic but now outdated *Southern Europe.* North Africa is included in *The Mediterranean lands* by D S Walker (third edition, Methuen, 1964), who has also written a standard regional text on Italy entitled simply *A geography of Italy* (second edition, Methuen, 1967). Both books contain short but well-chosen bibliographies. A detailed and well-documented study of a more limited region of this part of Europe is *The Western Mediterranean world: an introduction to its regional landscapes*, by J M Houston (Longmans, 1964), another example from Longmans' *Geographies for advanced study* series which includes several excellent regional texts. Moving to the east there is a further example from the same series in *Eastern Europe* by N J G Pounds (Longmans, 1969). About three-quarters of this typically detailed text is devoted to a country-by-country account, following several introductory chapters on the area as a whole. An alternative is *East-Central Europe: a geographical introduction to seven socialist states* by R H Osborne (Chatto and Windus, 1967), which is also regionally arranged but which excludes East Germany and is consistently less detailed throughout. A more recent study is *Eastern Europe: a geography of the Comecon countries* by R E H Mellor (Macmillan, 1975), a well-written text which progresses from the physical environment through history and demography to economics. *Central Europe: a regional and human geography* by A F A Mutton (second edition, Longmans, 1968) is less politically orientated, including discussion of Germany (East and West), Benelux, Austria, Switzerland, and Czechoslovakia. *Italy, Switzerland and Austria: a geographical study* by H Rees (Harrap, 1974) is a businesslike textbook providing a good introductory account of these countries.

Very valuable introductions to the literature of these less well documented areas of Europe are several guides edited by P L Horecky. *East Central Europe: a guide to basic publications,* and *Southeastern Europe: a guide to basic publications* were both published in 1969 by the University of Chicago Press, and both follow a similar pattern. Both contain sections on the whole area followed by detailed listings by country, and both are extensively annotated by experts. Each volume includes more than three thousand entries, arranged under subjects and fully indexed. The same editor has also prepared two volumes on the USSR, entitled *Basic Russian publications: an annotated bibliography on Russia and the Soviet Union* (University of Chicago Press, 1962), and *Russia and the Soviet Union: a bibliographic guide to western-language publications* (University of Chicago Press, 1965). The complementary nature of these sources is obvious from their scope as indicated in the titles. Both volumes contain specifically geographical sections prepared by Chauncy Harris, but as so often in these cases geographers will find much of interest under several headings. There is no shortage of general textbooks on Russia. *Russian land, Soviet people: a geographical approach to the*

USSR by J S Gregory (Harrap, 1968) is a full and detailed study with both topical and regional chapters. *Geography of the USSR* by P Lydolph (third edition, Wiley, 1977) is a more up-to-date text based on the nineteen statistical regions used by the Soviet authorities. *The Soviet Union: an economic geography* by R S Mathieson (Heinemann, 1975) is a profusely illustrated text with an extensive bibliography. Less complex works which provide a good basic introduction include *Geography of the USSR* by R E H Mellor (Macmillan, 1964); *The Soviet Union: people and regions*, D J M Hooson (University of London Press, 1966); *The Soviet Union*, G M Howe (MacDonald and Evans, 1968); and *A geography of the USSR*, J P Cole (Penguin, 1967).

Anglo-America

A good starting point for detailed literature searching on North America is provided by a bibliography prepared by the Library of Congress, entitled *A guide to the study of the United States of America* (1960). This lists over six thousand titles of 'representative books reflecting the development of American life and thought'. A second much slighter but more specifically geographical source is *A bibliographic guide to the economic regions of the United States,* by B J L Berry and T H Hankins (University of Chicago Department of Geography, 1963). Both bibliographies are well annotated, but of course neither is really up to date. *Current geographical publications* is the best of the indexing services for tracing more recent material, providing the most comprehensive coverage of American publications.

Basic standard regional geographies covering both Canada and the US include *Regional geography of Anglo-America* by C L White, E J Foscue, and T L McKnight (fourth edition, Prentice Hall, 1974) and *Anglo-America: a regional geography of the United States and Canada* by P F Griffin, R N Young, and R L Chatham (Methuen, 1963). Both are very fully illustrated and incorporate short bibliographies, and both are consistently regional in their arrangement. *The Anglo-American realm* by O P Starkey and J L Robinson (McGraw-Hill, 1969) is similar in appearance and in scope, but is primarily systematic in approach, with the exception of a separate section on Canada. Systematic and regional approaches are combined in *North America: a geography of Canada and the United States* by J H Paterson (fifth edition, Oxford U P, 1975), a readable study which like the previous works contains numerous illustrations. *North America: its countries and regions* by J W Watson (second edition, Longmans, 1968) contains the most substantial bibliography of this group.

Like Paterson it is partly regional and partly systematic in its organisation, and devotes particular attention to Canada. A shorter and very readable text is *The United States and Canada: a geographical study of regional problems* by W R Mead and E H Brown (revised edition, Hutchinson, 1970). This includes several chapters each on regions of the United States and Canada, in addition to introductory studies of the sub-continent as a whole.

There are few general regional geographies of the United States, although a useful and up-to-date treatment has been published recently in *A modern geography of the United States* by R Estall (Penguin, 1972). As its subtitle ('Aspects of life and economy') indicates this concentrates on the human and in particular on the economic aspects of the country. *The cultural geography of the United States* by W Zelinsky (Prentice Hall, 1973) is a further useful text for human geography, while a standard source on the physical side is *Regional geomorphology of the United States* by W D Thornbury (Wiley, 1965). This covers the country systematically, region-by-region. *The United States: a companion to American studies*, edited by D Welland (Methuen, 1974), consists of thirteen essays by various specialists on American geography, history, literature, politics, and society. *The American environment: perceptions and policies*, edited by J W Watson and T O'Riordan (Wiley, 1976), is not a comprehensive or systematic account but a study of selected problems, the link being provided by the perception theme featured in the sub-title. The same theme is pursued in *American environment*, edited by W R Mead (Athlone Press, 1974), in which three geographers explore changing attitudes to natural resources in the United States.

The most satisfactory and systematic studies of Canada are in several works associated with D F Putnam. These are *A regional geography of Canada* by D F Putnam and F P Kerr (revised edition, Dent, 1964); *Canada: a regional analysis* by D F and R G Putnam (Dent, 1971); and *Canadian regions: a geography of Canada* edited by D F Putnam (seventh edition, Dent, 1965). These are all strongly regional in their approach. A largely systematic discussion is found in a rather old but still interesting study entitled *Canada: a study of cool continental environments and their effect on British and French settlement* by G Taylor (third edition, 1957). *Canada: a geographical interpretation*, edited by J Warkentin (Methuen, 1968), is an important collection of papers by Canadian geographers, both regional and systematic. *Canada: a geographical perspective* by L E Hamelin (Wiley, 1973) is a translation of an interesting and readable study by a French-Canadian geographer, with a problem-orientated approach. *Studies in Canadian geography* edited by L Trotier (University of Toronto Press, 1973) consists of six short volumes, covering the country region by region. There is a short general rather than geographical guide to sources of information in *How to find out about Canada* by H C Campbell (Pergamon, 1967).

Latin America

The most valuable and basic bibliographical source on Latin America is the *Handbook of Latin American studies*, which has been published annually since 1935, first by Harvard U P and more recently by the University of Florida Press. It is not confined to geography and includes literature on a wide range of varied topics relevant to Latin American studies. Its coverage is selective, and the entries are annotated. Sources on the history and evolution of Latin America are listed in a valuable bibliography, *Latin American history: a guide to the literature in English* by R A Humphreys (Oxford U P, 1958, and reprinted). Other monograph bibliographies and useful lists within other works can be traced through *A bibliography of Latin American bibliographies*, compiled by A E Gropp (Scarecrow Press, 1968, with a supplement, 1971). This is a select list of over seven thousand items, arranged by subject with regional sub-divisions. The leading journal is *Revista geogrâfica* (Instituto Pan-Americano de Geografia e Historia, 1941-). This contains articles in Spanish, Portuguese, and English (summaries in English being included in the case of the former languages) and reviews, news and reports. There are two issues per year. Providing a rapid news service is the weekly journal *Latin America*, published by Latin American Newsletters Ltd. Short news reports are included relating to political, economic and social developments or events in any of the Latin American countries. Finally, mention should be made of the *South American handbook* (Trade and Travel Publications, annual). This is basically a travel guide, but incorporates a mass of statistical and other factual data for each country and is a very convenient source of reference.

Latin America by P E James (fourth edition, Odyssey Press, 1969) is perhaps the best known and most widely accepted general text on the area. The arrangement is by country, with a strong emphasis on Brazil. The text is well-illustrated throughout and there is a good bibliography, based like the text on the list of countries. More limited in scope is *Latin America: an economic and social geography* by J P Cole (second edition, Butterworths, 1975). The book is divided almost equally between a series of regional chapters and general sections relating to Latin

America as a whole and devoted to specific themes. A similar approach is followed in *Geography of Latin America: a regional analysis* by K F Webb (Prentice-Hall, 1972), a very concise and useful contribution in the *Foundations of world regional geography* series.

Latin America: a regional geography by G J Butland (third edition, Longmans, 1973) is an example of traditional methodology in regional geography. Each section is concerned with a particular area and describes its physical features before proceeding to human geography. Of similar scope are *Latin America: a geographical commentary* by I Pohl and J Zepp (Murray, 1966), and *South America: an economic and regional geography with an historic chapter* by E W Shanahan (eleventh edition, Methuen, 1963). The former is an adaptation by K E Webb of an original German text which contains short introductory studies of the area country by country. The latter is a well-established standard work originally published in 1927 and substantially revised in subsequent editions. *Latin America: New World, Third World*, by S Clissold (Pall Mall, 1972) is a more recent study which ranges broadly over historical, cultural, and political factors in an examination of the Latin American transition from new world to third world. A sound basic introduction is available in *South America*, edited by A Taylor (David and Charles, 1973), published in association with the American Geographical Society.

Social and economic factors are discussed in *Economies and societies in Latin America: a geographical interpretation* by P R Odell and D A Preston (Wiley, 1973) and in *Latin American development: a geographical perspective* by A Gilbert (Penguin, 1974). *Middle America: Its lands and peoples* by R C West and J P Augelli (second edition, Prentice-Hall, 1976) is a recent revision of an established standard reference on this area, with an emphasis on the importance of tradition and culture. An authoritative though not fully comprehensive study of the largest of the Latin American countries is available in *A geography of Brazilian development* by J D Henshall and R P Momsen (Bell, 1974). The Caribbean is covered in H Blume, *The Caribbean islands* (Longmans, 1974), translated from a German text originally published in 1968, and in a more recent study, *The West Indies* by F C Evans (Cambridge U P, 1974).

Africa

A useful starting point for the study of Africa is the bibliography *Africa: a bibliography of geography and related disciplines* by S H Bederman (third edition, Georgia State University, 1974). The title varies somewhat from previous editions. This is divided broadly into two regions, under which books and articles are listed alphabetically, and each section concludes with an index to the systematic branches of geography represented. At the end of the volume is an index of countries. A further valuable feature is the list of journals from which citations are taken, which includes a wide range of specialist periodicals devoted to African studies. This list includes such titles as *Africa report* (African-American Institute, nine issues per annum, 1956-); *African studies review*, formerly *African studies bulletin* (African Studies Association, three per annum, 1964-); *African affairs* (Royal African Society, quarterly, 1901-); *African studies* (Witwatersrand U P, quarterly, 1921-); and *Africa* (International African Institute, quarterly, 1928-); The first of these is primarily concerned with news and reports, providing a commentary and analysis of current affairs. The remainder are more academic, containing scholarly articles and book reviews. The list is a small selection of the more important among a great many other relevant journals. In view of this proliferation of journal literature, it is fortunate that there is a good English-language abstracting service in *African abstracts*. Published quarterly by the International African Institute, this contains short signed abstracts from a good selection of journals. In addition, the Institute publishes a quarterly *International African bibliography* which lists important new books and includes an index to articles before they appear in *African abstracts*. These two services used as a supplement to Bederman's bibliography provide a sound if not entirely comprehensive coverage of the literature. One of the yearbooks from Europa Publications, referred to above, covers *Africa south of the Sahara*. This includes separate chapters by experts on each country, incorporating directory material on the politics, government, economic situation, libraries, educational establishments and important personalities. It is a mine of useful information and a basic reference source.

A B Mountjoy and C Embleton, *Africa: a new geographical survey* (second edition, Hutchinson, 1967) is one of the better examples among the considerable number of general regional studies of Africa. It is a systematic and comprehensive account stressing both social and economic factors and the physical background. On the human aspects, this is obviously now in need of revision. A second good general text is *The geography of modern Africa* by W A Hance (second edition, Columbia U P, 1975), which has recently been revised and has an emphasis on economic affairs. *Africa* by L S Suggate (eleventh edition, Harrap, 1974) is a well-established textbook with a conventional approach, originally published in 1929 and more recently revised by several other scholars. Another up-dated text, now in its third edition, is *Africa: a study in tropical development* by L D Stamp and W T W Morgan (Wiley, 1972). This is about evenly divided between a topic-based approach to the whole continent and a series of more detailed regional chapters. The regional approach is abandoned altogether in the most recent of these substantial general geographies, *African survey* by A C G Best and H J de Blij (Wiley, 1977). In addition, there are several shorter introductions which contain less advanced treatment. These include *Africa* by N C Pollock (University of London Press, 1968), and *Africa and the islands* by R J Harrison Church, J I Clarke, P J H Clarke, and H J R Henderson (third edition, Longmans, 1972). Both volumes consist mostly of a regionally-organised description and analysis, with general introduction and conclusions relating to the continent as a whole.

African encyclopedia (Oxford U P, 1974) is a well-illustrated single volume encyclopedia, pitched at a fairly elementary level, which is nevertheless useful for reference purposes. There is also a very valuable collection of essays on several themes of interest to geographers, entitled *African studies since 1945: a tribute to Basil Davidson*, edited by C Fyfe (Longmans, 1976). This contains eighteen papers on history, archaeology, economy and so on. A more specifically geographical collection is *Africa in transition*, edited by B W Hodder and D R Harris (second edition, Methuen, 1972). The papers in this volume are original, and review the continent region by region. A select bibliography is appended in each case.

The emphasis of all of these volumes is on the pace of development and its impact in human terms, and must therefore incorporate some historical material. A fuller account which is both authoritative and readable is *Africa: history of a continent*, B Davidson (Macmillan, 1966). A number of important contemporary themes drawn from social and economic geography are featured in *Studies in emerging Africa* by N C Pollock (Butterworths, 1971). The organisation is systematic, and the selection of topics includes some unusual items, such as education and game conservation, in addition to the more predictable subjects. Similarly selective is *The geography of African affairs* by P Fordham (second edition, Penguin, 1968): the selection is intended to include 'such information as seems important for the understanding of current political and economic problems'. The result is a very concise and useful summary of the basic issues in the recent history of Africa.

Of course there are a great many more detailed systematic studies of particular topics, as there are more specific regional texts. Important examples of the latter include *West Africa: a study of the environment and of man's use of it* by R J Harrison Church (seventh edition, Longmans, 1974), and *East Africa* by W T W Morgan (Longmans, 1973), in the *Geographies for advanced study* series. Both are thorough and authoritative studies which usefully supplement the more general sources. A second important volume on West Africa is *West Africa* by W B Morgan and J C Pugh (Methuen, 1969), which is a well-documented and comprehensive text, arranged systematically. *A geography of Sub-Saharan Africa* by H J de Blij (Rand-McNally, 1964) is a basic introductory treatment of most of the continent, arranged under countries. It does not attempt factual completeness in every case, but each chapter emphasises a particular aspect of the geography of the country in question.

Asia

Asia is a vast continent incorporating widely differing physical and cultural regions and there is a great deal of confusion in the terminology generally used to indicate divisions of the continent. The designations used to identify these regions in the literature are not always consistent or generally accepted, and the continent is therefore discussed here as a whole. In fact very few of the works cited actually relate to the entire continent. *Asia: a regional and economic geography* by L D Stamp (twelfth edition, Methuen, 1967) is a thorough survey which has been established as a standard text for several years. Its arrangement is based on a regional approach. A second basic general study is *Asia's lands and peoples: a geography of one third of the earth and two thirds of its people* by G B Cressey (third edition, McGraw-Hill, 1963). This again is regional, following a relatively short statement on the continent as a whole, and is fully illustrated. Also written on a continental scale is *The changing map of Asia: a political geography*, edited by W G East, O H K Spate, and C A Fisher (fifth edition, Methuen, 1971). In this volume specialists analyse political patterns in the continent region by region. Each section includes a bibliography.

Major standard works dealing with the whole of Asia except for the Middle East and Asiatic Russia include H Robinson, *Monsoon Asia: a geographical survey* (third edition, McDonald and Evans, 1976) and *Asia, east by south: a cultural geography* by J E Spencer and W L Thomas (second edition, Wiley, 1971). The latter title includes a short but well selected bibliography and a useful digest of up-to-date statistical information. About half the text comprises a discussion of the area under systematic topics, which consider the physical basis as well as the human features of the continent. *Bibliography of Asian studies* (Association for Asian Studies, 1941-) formerly known as *Far Eastern bibliography*, is an annual publication which reviews literature on the same area, ie Asia excluding the Middle East. Each issue is published as one number in the *Journal of Asian studies*. The bibliography is regionally-organised, with systematic subdivisions, and includes entries for books, articles, theses and other materials, without annotation. While this is the most valuable source of current literature, a number of additional bibliographies and other reference sources are listed in *Asia: a selected and annotated guide to reference works* by G R Nunn (MIT, 1971).

The Middle East and North Africa is an annual directory and reference book similar in content and arrangement to its companion volume on *Africa south of the Sahara* already described. It features surveys of each country and a series of essays on matters of general concern as well as directory-style information and statistics. In addition there are several standard regional geographies, including *South-west Asia* by W C Brice (University of London Press, 1967) and *The Middle East: a social geography*, S H Longrigg and J Jankowski (second edition, Duckworth, 1970), both of which are sound and reliable texts. Probably the best all-round treatment, however, is *The Middle East: a physical, social and regional geography* by W B Fisher (sixth edition, Methuen, 1971). This example from *Methuen's advanced goegraphies* series has been widely accepted as a balanced and definitive study since its original publication in 1950, and subsequent editions have brought it fully up-to-date. A more recently published text, which also provides a comprehensive and well-written account of the area and could well become a standard reference, is *The Middle East: a geographical study* by P Beaumont, G H Blake, and J M Wagstaff (Wiley, 1976). A useful starting-point for tracing a more detailed literature is available in *The contemporary Middle East: a selective and annotated bibliography* by G N Atiyeh (G K Hall, 1975), and *The Near East (South-east Asia and North Africa): a bibliographic study* by J Zuwiyya (Scarecrow, 1973). A more comprehensive listing of older material is *Bibliography of Southwestern Asia* by H Field. This was published by the University of Miami Press, 1953-1962, in seven volumes each divided into three sections, on anthropogeography, zoology and botany. There are two sets of subject indexes, for the first five and the last two volumes respectively. Regular notification of more recent literature is found in the *Middle East journal* (quarterly, Middle East Institute, 1947-), which also contains many pertinent articles of a high standard.

The Far East and Australasia is yet another yearbook from Europa Publications, again providing very similar information and in a similar format to those mentioned above. Areas covered include South Asia, South-east Asia, and East Asia, as well as Australasia and the Pacific.

Important general texts on South and South-east Asia include *India, Pakistan, and Ceylon: the regions,* by O H K Spate and A T A Learmonth (Methuen, 1972); *South-east Asia* by E H G Dobby (eleventh edition, University of London Press, 1973); *East Asia* by A Kolb (Methuen, 1971, translated by C A M Sym from the German edition of 1963); and *South-east Asia: a social, economic and political geography*, C A Fisher (second edition, Methuen, 1966). The first of these follows a traditional pattern, proceeding from physical to human geography and finally providing a logical regional framework for the area. Kolb's volume, on China, Japan, Korea, and Vietnam, is a comprehensive and detailed study containing many excellent maps and emphasising the cultural background to current developments. Fisher's *South-east Asia* is a readable and established text which includes both topical and regional discussion, providing a good general introduction. *Man and land in the Far East* by P Gourou (Longmans, 1975) is a translation of a French text originally published in 1940, and now up-dated by means of a supplementary chapter.

Useful sources of further literature include *South Asia: a systematic geographic bibliography* by B L Sukhal (Scarecrow, 1974), containing more than ten thousand titles, with an emphasis on English language material; and *South-east Asia: a critical bibliography* by K G Tregonning (University of Arizona, 1969). There are of course many more major geographical works devoted to this part of the world, including many individual country studies; while it would be inappropriate to detail these here, reference must be made to *Japan: a geography*, G T Trewartha (Methuen, 1965), and to *Geography of China* by T R Tregear (University of London Press, 1965), both of which are standard works on two of the most important countries in the recent history of Asia, and both could now do with some revision. Further important and more recent sources on China are *China: an integrated study* by A Cotterell and D Morgan (Harrap, 1975), in which a geographer and a historian combine successfully to produce a most useful general text; and *China: a handbook*, edited by Yuan-Li Wu (David and Charles, 1973), a reference book mainly by Chinese and American authors containing short chapters on various aspects of the country, translations of key documents, and much useful statistical and other factual data.

Australasia and the Pacific

A Pacific bibliography: printed matter relating to the native peoples of Polynesia, Melanesia and Micronesia by C R H Taylor (second edition, Oxford U P, 1965) is a good starting-point for literature on the Pacific area up to 1960, including New Zealand but excluding Australia. The arrangement is regional, with topical subsections, and the materials listed comprise a wide range of publications in differing formats and languages, and including additional bibliographic sources. Unfortunately there are no annotations. Useful recent literature is reviewed by the journals of active geographical societies in Australia and New Zealand, and by the semi-annual *Pacific viewpoint* (1960-), published by the Department of Geography at the Victoria University of Wellington. Further material on Australia may be traced through *Australian bibliography: a guide to printed sources of information* by D H Borchardt (third edition, Pergamon, 1976). This describes major libraries and catalogues, reference works, and subject and specialised area bibliographies.

There are several general texts covering the area as a whole, in addition to more specialised regional sources. *Australia, New Zealand, and the Southwest Pacific*, K W Robinson (second edition, University of London Press, 1968) is very largely regional in its arrangement but includes several general introductory chapters covering major themes. The relative priority of the areas discussed conforms to the order established by the title. The same feature is evident to some extent in *Southwest Pacific: a geography of Australia, New Zealand, and their Pacific Island neighbours* by K B Cumberland (fourth edition, Methuen, 1968), although an attempt is made to create an awareness of the significance of the geographical proximity of these culturally distinct communities. While this is a stated aim of the book, the arrangement does not permit the theme the emphasis which it obtains in *Dilemmas down under: Australia and the Southwest Pacific* by A J Rose (Van Nostrand, 1966). This is a very concise study which examines Australia's situation and the problems arising from her position in the Pacific. A fuller account of the islands is found in *Geography of the Pacific*, edited by O W Freeman (Wiley, 1951, and more recent reprints). This is a collection of some nineteen papers, mostly concerned with particular island groups but including also several general essays. There is less emphasis on Australia and New Zealand, and the volume is particularly strong on physical and historical information; the economic and human data is of course rather dated.

The human aspects of Australian geography are discussed in a very readable volume, *Australia* by O H K Spate (Benn, 1968). The book is divided into four sections, on history, economy, polity, and society, and there is a short note on sources of further information. *Australia: a study of warm environments and their effect on British settlement* by G Taylor (seventh edition, Methuen, 1959) extends the discussion to include the physical background but the analysis is suggestive of deterministic theories which are not generally accepted in modern geography. A short but balanced introduction is provided in *Australia's corner of the world* by T L McKnight (Prentice-Hall, 1970), which also includes a chapter on New Zealand. *Studies in Australian geography*, edited by G H Dury and M I Logan (Heinemann, 1968) is an important collection of essays which range over a variety of topics although, as the introduction states, they are 'representative rather than comprehensive'. *Australia* by R L Heathcote (Longmans, 1975) is an original and interesting study with an historical emphasis and some discussion of differing perceptions of the Australian environment over the last two hundred years. Standard regional geographies of New Zealand include *New Zealand* by K B Cumberland and J S Whitelaw (Longmans, 1970), in the *World's landscapes* series, and at a more elementary level, *New Zealand geography* by R D Mayhill and H G Bawden (Blackwood and Janet Paul, 1966). *Society and environment in New Zealand,* edited by R J Johnston for the New Zealand Geographical Society (Whitcombe and Tombs, 1974) is an important collection of papers written for the International Geographical Union Regional Congress held in New Zealand in 1974. Papua and New Guinea are covered in *The geography of Papua and New Guinea* by D Howlett (Belson, 1973), and substantial general books on the Pacific islands include *Melanesia: a geographical interpretation of an island world* by H C Brookfield and D Hart (Methuen, 1971, in the *Advanced geographies* series), and *Island populations of the Pacific*, N McArthur (Australian National University Press, 1967). *The Philippine Island world: a physical, cultural, and regional geography* by F L Wernstedt and J E Spencer (University of California Press, 1967) is a thorough and systematic review of this area.

Polar regions

Literature on the Polar regions is well covered in several journals and bibliographic serials. The Arctic Institute of North America prepares an *Arctic bibliography*, which has been published since 1953, usually annually. This is a comprehensive list and contains abstracts. Originally published by the US Government Printing Office, more recent issues come from McGill University Press. *Antarctic bibliography* is a similar publication prepared since 1965 by the Library of Congress under the auspices of the Office of Polar Programs, National Science Foundation. Again the frequency is normally annual. Both sources contain materials in a wide range of languages. These rather infrequent lists can be supplemented by journals such as *Arctic* (quarterly, 1948-), the journal of the Arctic Institute of North America; *Antarctic journal of the United States* (bi-monthly, 1966-); and *Polar record*, published three times a year by the Scott Polar Research Institute (1931-) and listing recent literature relating to both poles.

INDEX

Most of the entries in this index are for personal and corporate authors and for the titles of serials and periodicals; these are listed comprehensively. Entries under forms of literature are confined to descriptions of the character and functions of these forms in general, and to examples relating to geography as a whole; further items will be found under their individual authors or titles, or may be traced by subject. The subject indexing is limited to broad terms only.